SIGNIFICANT LIQUID STRUCTURES

SIGNIFICANT LIQUID STRUCTURES

HENRY EYRING
Professor of Chemistry
University of Utah
Salt Lake City, Utah

MU SHIK JHON
Assistant Professor of Chemistry
University of Virginia
Charlottesville, Virginia

 John Wiley & Sons, Inc., New York · London · Sydney · Toronto

PREFACE

An acceptable model of the liquid state should suggest confirmatory experiments as well as lead to quantitative calculations of thermodynamic and transport properties. Significant structure theory is such a model. Two of the most striking characteristics accompanying melting are a marked volume change and a marked increase in fluidity. The fact that X-ray diffraction indicates little or no change in the nearest neighbor distance with melting suggests a retention of small regions of solid-like structure interspersed with loose, gas-like regions where molecules unsupported by neighbors, fall freely for short distances. An instantaneous photograph would, accordingly, reveal an intimate mixture of solid-like and gas-like degrees of freedom. That this heterogeneity is fine grained is attested to by the fact that liquids can be prepared in which there are no solid-like or gas-like nuclei large enough to form foci which prevent the appearance of supercooling and the bumping of liquids. In order that loose regions can exist in a fluid, there must be a local balance between the kinetic energy density tending to make a region expand and the potential energy density tending to make these regions collapse. This balance is necessary for a phase to be dynamically stable and accounts for the existence of a melting temperature. This dynamical stability of a phase can persist into a region of thermodynamic instability, i.e., the supercooled liquid region, since the speed of appearance of nuclei about which the stable phase can grow is the rate-determining step.

That one should expect a more or less faithful mirroring of molecules in the vapor phase in the form of vacancies moving in a solid-like sea follows from the fact that the energy to form a vacancy is equal to the energy of vaporization. One should also expect that "fluidized vacancies" will collide and aggregate in a fluid (but not in the solid) much as molecules perform in the vapor state. The development of this model, somewhat reminiscent of the concept of electrons in the conductance band mirrored as holes in the valence band, is the theme of this book. Obviously the nature of the model guaranteed partial success, but the ease of application and the faithful characterization of the most varied thermodynamic and transport properties could scarcely have been anticipated. There seems to be no liquid which cannot be usefully examined using this model.

Detailed justification of the model of significant structure theory from first principles is difficult inasmuch as a vacancy is moved about by the cooperative action of all its neighbors, their motion being correlated in such a way as to transform three degrees of freedom which would otherwise be solid-like vibrations into three translational degrees of freedom characteristic of a gas molecule, although no single gas molecule having this property may exist. Solving such a many-body problem exactly has much of the complexity of the exact phase integral for a liquid. Our model starts with what we conjecture is the correct solution of this difficult normal coordinate problem.

By gathering together the material in this rapidly developing field a scattered literature becomes readily available and an individual assessment of the applicability of the model can be made. If this effort advances in some degree our understanding of the liquid state the authors will feel well repaid.

HENRY EYRING

MU SHIK JHON

March 1969

CONTENTS

chapter 1

FORMAL THEORY OF LIQUIDS

1.1 INTRODUCTION

A theory of the liquid state is one of the most important problems of statistical physics and is essential for the further development of many branches of physics and physical chemistry. However, due to the characteristics of the liquid state, i.e., the strong interaction of particles and their state of disorder, the theoretical analysis has lagged far behind theories of the gaseous and crystalline states. Over the years there have been two kinds of developments in the theory. One, the formal approach, pioneered by Mayer and Kirkwood among others, provides the aesthetic pleasure of initial rigor but, at some stage of the development, approximations have always crept in to make the mathematics tractable. When such difficulties are overcome, a noteworthy advance will have been achieved. In this chapter, we will describe the method of the radial distribution function which is based on the formal approach. The alternative procedure is the model approach, in which one visualizes a physical model of the liquid, translates the picture into mathematical language, i.e., a partition function, and then calculates the properties of the liquid. If the model is reasonable and the theory predicts the experimental values with sufficient faithfulness, one assumes that the model is an adequate description of reality. This approach will be described in later chapters.

1.2 THE METHOD OF THE PAIR DISTRIBUTION FUNCTION

The formal approach to an understanding of the properties of fluids is based on attempts to calculate the radial distribution function, $g(r)$. For fluids whose molecules interact with central additive forces, the knowledge of $g(r)$ gives us the thermodynamic properties through the use of the equation of state and the total energy expressions.

To obtain the ordinary equation of state, we make use of the virial theorem.

1

If the potential is assumed to be due to the additive contributions of all pairs of molecules, i.e.,

$$H = \sum_i \frac{p_i^2}{2m} + \sum_{i>j} U(r_{ij}) \tag{1.1}$$

then, using the virial theorem the equation of state becomes

$$PV = NkT - \frac{1}{3} \overline{\sum_{i>j} r_{ij} \frac{\partial U}{\partial r_{ij}}} \tag{1.2}$$

If the time average in Eq. (1.2) is replaced by an average over a canonical ensemble, and integrated over the momenta, we obtain

$$PV = NkT - \frac{1}{3N! \, Q_N} \int \sum_{i>j} r_{ij} \frac{\partial U}{\partial r_{ij}} \exp \left\{ -\frac{1}{kT} \sum_{i>j} U(r_{ij}) \right\} dr_1 \cdots dr_N \tag{1.3}$$

where Q_N, the configurational partition function, is given by

$$Q_N = \frac{1}{N!} \int \exp \left\{ -\frac{1}{kT} \sum_{i>j} U(r_{ij}) \right\} dr_1 \cdots dr_N \tag{1.4}$$

Here, it is convenient to define the molecular-pair distribution function $n^{(2)}(r_1, r_2)$ which is the probability of finding an arbitrary pair of molecules in the configuration r_1, r_2,

$$n^{(2)}(r_1, r_2) = n^{(2)}(r_{12}) = \frac{N!}{(N-2)!} \frac{\int e^{-U/kT} dr_3 \cdots dr_N}{\int e^{-U/kT} dr_1 \cdots dr_N} \tag{1.5}$$

Similarly, other distribution functions $n^{(h)}$ can be defined

$$n^{(h)}(r_1, r_2, \ldots, r_h) = \frac{N!}{(N-h)!} \frac{\int e^{-U/kT} dr_{h+1} \cdots dr_N}{\int e^{-U/kT} dr_1 \cdots dr_N} \tag{1.6}$$

Integration of the second term on the right-hand side of Eq. (1.3) over the coordinates of all molecules except i and j leads to $\frac{1}{2}N(N-1)$ terms of the form

$$\iint r_{ij} \frac{\partial U}{\partial r_{ij}} n^{(2)}(r_{ij}) \, dr_i \, dr_j \tag{1.7}$$

Hence, the equation of state becomes

$$PV = NkT - \frac{1}{6} \iint n^{(2)}(r_{12}) r_{12} \frac{\partial U}{\partial r_{12}} \, dr_1 \, dr_2 \tag{1.8}$$

The internal energy can be similarly related to the molecular pair distribution function:

$$E = \frac{1}{N! f_N} \int \left\{ \sum_i \frac{p_i^2}{2m} + \sum_{i>j} U(r_{ij}) \right\} e^{-H/kT} dp_1 \cdots dp_N \, dr_1 \cdots dr_N \tag{1.9}$$

where f_N is the partition function of the system of N particles. Using the principle of equipartition of energy in a manner analogous to that used in Eq. (1.8), we obtain

$$E = \tfrac{3}{2}NkT + \tfrac{1}{2} \iint n^{(2)}(r_{12})U(r_{12}) \, dr_1 \, dr_2 \qquad (1.10)$$

Hence, the evaluation of the pair distribution function predicts the thermodynamic properties of a liquid. Since it is related to the radial distribution function $g(r)$ by

$$n^{(2)}(r_{12}) = \frac{N(N-1)}{V^2} g(r_{12}) \qquad (1.11)$$

immediate calculations of E and P might be thought to be possible from the value of $g(r_{12})$ obtained from diffraction experiments. Unfortunately, the accuracy with which $g(r_{12})$ is at present known is quite inadequate for this purpose. One analytic procedure is to develop $n^{(2)}(r)$, and thus P or E, in a power series in the density. However, this method, although rigorous, suffers from the slow convergence of the series and also the virial coefficients are very hard to obtain.

Here, we shall only be concerned with another method in which an integral equation is obtained for $n^{(2)}(r)$. Actually there are various expressions for the integral equations of $n^{(2)}(r)$ obtained by Born and Green [1],* Yvon [2], Kirkwood [3], and Bogoliubov [4]. These expressions are basically equivalent and can be called the B.B.G.Y.K. equations. We will follow the procedure due to Yvon and to Born and Green, the B.G.Y. equation.

For simplicity, let us consider a pair of molecules, 1 and 2. When we vary the position of molecule 1 by a small displacement dr_1 while keeping molecule 2 fixed, the effect of this change on $n^{(2)}(r_{12})$ is

$$\frac{\partial n^{(2)}(r_{12})}{\partial r_1} = \frac{1}{(N-2)! \, Q_N} \frac{\partial}{\partial r_1} \int \exp \frac{-U}{kT} \, dr_3 \cdots dr_N \qquad (1.12)$$

Now, U depends on r_1 (as do all the other coordinates) but the limits of integration do not. We therefore differentiate under the integral sign and obtain

$$\frac{\partial n^{(2)}(r_{12})}{\partial r_1} = \frac{1}{(N-2)! \, Q_N} \int \exp\left(-\frac{U}{kT}\right)\left(-\frac{1}{kT}\frac{\partial U}{\partial r_1}\right) dr_3 \cdots dr_N \qquad (1.13)$$

To evaluate $\partial U/\partial r_1$, let us write some of the terms on which it depends:

$$U = U_{12} + U_{13} + \cdots + U_{1N} + U_{23} + U_{24} + \cdots \qquad (1.14)$$

* References appear at the end of each chapter.

then

$$\frac{\partial U}{\partial r_1} = \frac{\partial U_{12}}{\partial r_1} + \sum_{k=3}^{N} \frac{\partial U_{1k}}{\partial r_1} \qquad (1.15)$$

Substituting Eq. (1.15) into Eq. (1.13), we obtain

$$\frac{\partial n^{(2)}(r_{12})}{\partial r_1} = \frac{-1}{kT(N-2)!\,Q_N} \int \frac{\partial U_{12}}{\partial r_1} \exp\left(-\frac{U}{kT}\right) dr_3 \cdots dr_N$$

$$- \frac{1}{kT(N-2)!\,Q_N} \int \sum_{k=3}^{N} \frac{\partial U_{1k}}{\partial r_1} \exp\left(-\frac{U}{kT}\right) dr_3 \cdots dr_N \quad (1.16)$$

The quantity $\partial U_{12}/\partial r_1$ in the first term on the right-hand side does not depend on the variables of integration and thus can be treated as a constant and placed outside the integral. Using the relation in Eq. (1.5) this term is just

$$- \frac{1}{kT} \frac{\partial U_{12}}{\partial r_1} n^{(2)}(r_{12})$$

The second expression on the right-hand side is the sum of $(N-2)$ terms, one for each value of k from 3 to N. The first of these terms is

$$- \frac{1}{kT(N-2)!\,Q_N} \int \frac{\partial U_{13}}{\partial r_1} \exp\left(-\frac{U}{kT}\right) dr_3 \cdots dr_N \qquad (1.17)$$

which can be rewritten:

$$- \frac{1}{kT(N-2)!\,Q_N} \int_{r_3} \frac{\partial U_{13}}{\partial r_1} \left[\int \exp\left(-\frac{U}{kT}\right) dr_4 \cdots dr_N\right] dr_3 \qquad (1.18)$$

where the integrations in the square brackets are to be performed first. Now, if we put $h = 3$ in Eq. (1.6), we see that the square bracket in Eq. (1.18) is equal to $n^{(3)}(r_1, r_2, r_3)(N-3)!\,Q_N$. Thus, Eq. (1.18) simplifies to

$$- \frac{1}{kT(N-2)} \int \frac{\partial U_{13}}{\partial r_1} n^{(3)}(r_1, r_2, r_3)\, dr_3 \qquad (1.19)$$

In a similar manner, the term corresponding to $k = 4$ of the summation in Eq. (1.16) is equivalent to

$$- \frac{1}{kT(N-2)} \int \frac{\partial U_{14}}{\partial r_1} n^{(3)}(r_1, r_2, r_3)\, dr_4 \qquad (1.20)$$

We can readily see that the values of the integrals in both Eq. (1.19) and Eq. (1.20) are the same. Therefore, all the terms in the summation in Eq. (1.16) have the same value as Eq. (1.19). There is a factor $(N-2)$ in the denominator of Eq. (1.19). Thus, Eq. (1.16) finally becomes

$$\frac{\partial n^{(2)}(r_1, r_2)}{\partial r_1} = -\frac{1}{kT} n^{(2)}(r_1, r_2)\frac{\partial U_{12}}{\partial r_1} - \frac{1}{kT}\int \frac{\partial U_{13}}{\partial r_1} n^{(3)}(r_1, r_2, r_3)\, dr_3$$

$$(1.21)$$

This equation means that the change in the pair distribution function $n^{(2)}(r_1, r_2)$ due to a small displacement of molecule 1, keeping molecule 2 fixed, is the sum of two parts: The first comes from the direct force $-\partial U_{12}/\partial r_1$ exerted by molecule 2 on molecule 1; the second arises from the forces exerted by all the other molecules averaged over all their possible configurations. To derive a reasonably simple equation for $n^{(2)}(r_1, r_2)$, we have obtained an expression of the unknown triplet function which appeared in its derivation. By an analogous calculation, we may find an expression for $\partial n^{(3)}(r_1, r_2, r_3)/\partial r_1$ but this leads to an equation involving the next higher function, $n^{(4)}(r_1, r_2, r_3, r_4)$, and so on. In order to solve the problem, we must break the chain of linked equations. This has been done by abandoning the rigorous attempt to calculate the pair distribution and introducing a simplifying assumption.

In 1935 Kirkwood made the following assumption, known as the superposition approximation. He proposed to approximate the triplet function by a product of pair functions

$$n^{(3)}(r_1, r_2, r_3) = \frac{[n^{(2)}(r_1, r_2) n^{(2)}(r_2, r_3) n^{(2)}(r_3, r_1)]}{\rho_0} \qquad (1.22)$$

Here ρ_0 is the number-density of the molecules, i.e., the probability that any volume element dv_1 contains a molecule is $\rho_0\, dV_1$. When we use the superposition approximation for $n^{(3)}(r_1, r_2, r_3)$ in Eq. (1.21), we can take $n^{(2)}(r_1, r_2)$ outside the integral to give

$$\frac{\partial n^{(2)}(r_1, r_2)}{\partial r_1} = -\frac{n^{(2)}(r_1, r_2)}{kT}\frac{\partial U_{12}}{\partial r_1}$$

$$-\frac{n^{(2)}(r_1, r_2)}{kT\rho_0^3}\int \frac{\partial U_{13}}{\partial r_1} n^{(2)}(r_2, r_3) n^{(2)}(r_3, r_1)\, dr_3 \qquad (1.23)$$

This equation contains only pair distributions, so we have broken the chain of equations which leads to the higher order functions. Born and Green [1] and Yvon [2] have shown by means of straightforward but lengthy manipulations that Eq. (1.23) can be integrated over r_1, giving

$$kT \ln \frac{n^{(2)}(r_{12})}{\rho_0^2} = -U_{12} + 2\pi\rho_0 \int_0^\infty \int_{r-r_{13}}^{r+r_{13}} \frac{\partial U_{13}}{\partial r_{13}} \frac{n^{(2)}(r_{13})}{\rho_0^2}\left(\frac{n^{(2)}(r_{23})}{\rho_0^2} - 1\right)$$

$$\times \left[\frac{r_{13}^2 - (r_{23} - r_{12})^2}{2r}\right] r_{23}\, dr_{23}\, dr_{13} \qquad (1.24)$$

where $u(-r) = u(r)$ and $n^{(2)}(-r) = n^{(2)}(r)$ by definition.

Kirkwood [3], by a rather different approach, also obtained this equation. In his procedure, the superposition approximation was inserted but the final form of Kirkwood's equation differs from the B.G.Y. equation. The B.G.Y. integral equation, the Kirkwood, and the Bogoliubov equations, all are accordingly approximate.

If $n^{(2)}$ is expanded in a power series in the density, it is found [5] that the resulting equation of state is correct for the second and third virial coefficients but not for the higher coefficients. This is due to the inadequacies of the superposition assumptions. Modifications of the superposition approximation have been studied [6] with improved results for the fourth virial coefficient. Kirkwood and co-workers [7] obtained the solutions of the B.G.Y. equation for the case of (a) a system of rigid spheres, (b) the Lennard-Jones (L.J.) potential with a hard sphere "cut off" near the origin, and finally (c) the unmodified L.J. potential. For all cases the calculated $g(r)$ reproduced the broad features of the diffraction data. They show reasonable agreement with experimental data for the thermodynamic properties. However, at higher densities, the results are only moderately good. This again indicates the inadequacies of the superposition principle at high densities. More recently it was possible to obtain an exact integral equation for the pair distribution function [8–10]; the assumptions made subsequently appear to be better than the superposition approximation because they are self-consistent and are introduced after the derivation of the integral equation. However, detailed numerical results have not yet been obtained.

1.3 THE METHOD OF COLLECTIVE VARIABLES

A radically different theory of liquids has been proposed known as the method of collective variables [11]. This method was originally applied by Bohm and Pines [12], not to liquids, but to problems such as the interaction of charged particles in ionized gases. In such a gas (a plasma) system, the electrostatic potential energy of a pair of charged particles varies rather slowly with the distance of separation, i.e., as the inverse first power. However, in liquids, the potential energy varies much more rapidly, approximately as the inverse sixth power at large distances, and the application of the collective variable method is accordingly more difficult. In 1958, Percus and Yevick [11] applied the method to liquids and, after a lengthy analysis, were able to formulate an integral equation, the P.Y. equation, for the radial distribution function which is very different from the previous B.B.G.Y.K. equation. Their procedure will be outlined briefly.

It is well known that the fundamental difficulty in calculating the partition function arises from the intractable properties of the integrations over the molecular coordinates in the configuration integral for the liquid. Because any one molecule is in strong interaction with many other molecules, the

Hamiltonian function, which includes the potential energy of the molecules, cannot be separated into parts that depend on the coordinates of individual molecules only.

According to the method of collective variables, the Hamiltonian can be transformed so that this separation is approximately realized. Suppose we divide the space occupied by a liquid into small volume elements (boxes) whose sides are a few angstrom units in length. The number of molecules in any box will fluctuate in time about a mean value as the molecules move about. Because of the intermolecular forces, the fluctuations in any one box will influence those in neighboring boxes. As a result, the density variations will tend to be propagated through the liquid, at least for short distances.

Now, we consider that the propagation of a density fluctuation is analogous to the familiar phenomena of the transmission of a sound wave, i.e., alternations of compressions and rarefactions traveling through the medium. We seek to express the microscopic fluctuations in liquids in terms of acoustic waves that constantly traverse the liquid in all directions. These waves are another way of representing the unceasing molecular motion. A single simple harmonic wave would not be an adequate expression of course, since this would correspond to a regular long-range variation in the number-density, and in liquids long-range order is absent. Hence, we must describe the random microscopic density fluctuations in a liquid as a superposition of all possible waves. The motion of an elastic string fixed at the points $x = 0$ and $x = L$ provides a simple example of the superposition of standing waves. If the string is set in motion, the displacement at any point will vary in a complex manner which can be described by the superposition of a large number of simple harmonic waves. According to the familiar Fourier analysis, the allowed frequencies are of the form $\sin 2\pi x/\lambda$, where $\lambda = 2L/n$ and n is an integer. We can use this analogy in describing a liquid according to the collective variable method.

Let us consider a liquid system of N particles moving in a line lying between $x = 0$ and $x = L$. In the ordinary statistical mechanical theory of such a one-dimensional liquid, we describe the instantaneous configurations in terms of the particle coordinates, $x_1, x_2, \ldots, x_n$. On the other hand, each collective variable depends on all the x_i's. For example, the first collective variable is defined as

$$q_1 = \sum_{i=1}^{N} \sin 2\pi x_i/\lambda_1$$

where the wave length, λ_1, is one of the waves that forms the system. In general, the collective coordinate has the form

$$q_n = \sum_{i=1}^{N} \sin 2\pi x_i/\lambda_n \tag{1.25}$$

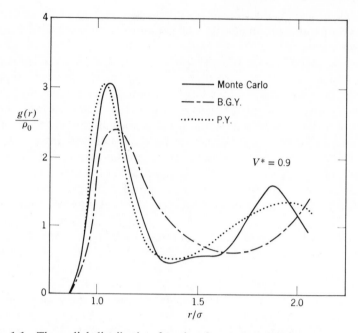

Figure 1-1   The radial distribution function for an L.J. (6-12) fluid at $T^* = 2.74$ $V^* = 0.9$ according to three theories. (after Broyles *et al.* [13])

To describe a configuration by the usual method, we need N coordinates, x_1 to x_n, whereas using the collective variable method, we need only the N collective coordinates, q_1 to q_n. For a real three-dimensional fluid, we need $3N$ collective coordinates and consider the liquid to be confined to a cube of length L on a side.

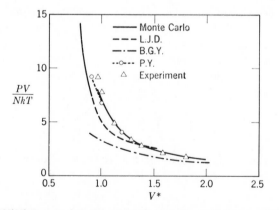

Figure 1-2 PV/NkT as a function of V^* at $T^* = 2.74$ for an L.J. (6-12) fluid, according to four theories. (after Broyles *et al.* [13])

The next step is to reformulate the Hamiltonian function. The greatest difficulty is the transformation of the potential energy U which is usually represented by the L.J. (6-12) potential. This difficulty is connected with the fact that the minimum allowed wavelength is $2L/N$, while the L.J. (6-12) potential varies rapidly over distances comparable with this wavelength. Actually, the transformation involves complex mathematical procedures whose justification is still open to question. Finally, the Hamiltonian is separated into a sum of terms each of which depends only on one collective variable. Using this Hamiltonian, we can calculate the thermodynamic properties. The following is the integral equation obtained by Percus and Yevick [11]:

$$e^{U(r)/kT}g(r) = 1 + \rho_0 \int (1 + e^{U(r_1)/kT} g(r_1)[g(|r - r_1|) - 1] \, dr_1 \quad (1.26)$$

The solution of the P.Y. integral equation has been obtained by Broyles [13] subject to the same conditions used for the B.G.Y. equation. The radial distribution function and thermodynamic properties are compared in Figures 1-1, 2, and 3. It can easily be seen that the P.Y. equation is a great improvement over the B.G.Y. equation. The first peak in the radial distribu-

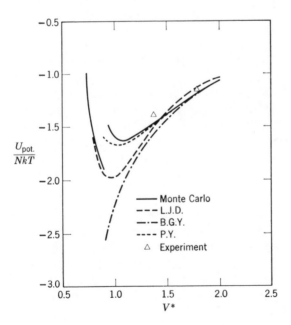

Figure 1-3   The potential energy, in molecular reduced units, of an L.J. (6-12) fluid at $T^* = 2.74$ as a function of V^*, according to four theories. (after Broyles *et al.* [13])

tion function is in very good agreement with the Monte Carlo values (Figure 1-1). At large distances, less satisfactory results are obtained, although they are still better than those given by the B.G.Y. equation. The P.Y. pressures (Figure 1-2) and configurational internal energy (Figure 1-3) are much better than those given by the B.G.Y. equation.

No doubt, the collective variable approach marks a significant advance in the formal theory of liquids. Final judgment, however, must be deferred until more numerical calculations have been made and compared with experiment. In the low density region of fluids, the method is of less value, since the molecular interactions are restricted to binary collisions. For this situation, pair correlation methods of adequate accuracy are available. Recently, an analytical solution of the P.Y. equation was obtained by Wertheim [14] and Thiele [15].

1.4   THE SCALED PARTICLE THEORY

To conclude this chapter, we will describe briefly the scaled particle theory developed by Reiss *et al.* [16]. This method provides an evaluation of the new distribution function $G(r)$ which measures the density of rigid sphere molecules in contact with a rigid sphere solute of arbitrary size. This theory is particularly suitable for the hard sphere potential but has also proved useful in predicting the properties of real fluids [17]. Another advantage of this theory is that the derivation of the equation of state follows readily from intuitive considerations. The basic idea is that there are discontinuities in the number of molecules that can occupy a void in a liquid as a function of the void volume, i.e., the volume when small can hold only one molecule and this discontinuously jumps to two molecules when the volume of the void exceeds a critical value, etc. This method is developed in terms of the nearest-neighbor distribution function, rather than in terms of the pair correlation function. For rigid spheres, the pressure equation can be expressed:

$$\frac{P}{\rho_0 kT} = 1 + \tfrac{2}{3}\pi\rho_0\sigma^3 g(\sigma) \qquad (1.27)$$

Here σ denotes the diameter of the rigid sphere and the rest of the notation has been defined previously. Thus, to obtain the equation of state, only $g(\sigma)$ is evaluated. If we define $P_0(r)$ as the probability of finding a spherical hole of at least radius r centered about some specified point in the liquid, then $-dP_0(r)/dr$ is the probability of finding a hole with radius lying between r and $r + dr$. Thus

$$-\frac{dP_0(r)}{dr} = P_0(r)4\pi r^2\rho_0\,G(r) \qquad (1.28)$$

where $\rho_0 G(r)$ is the average density of hard sphere centers in contact with the boundary of the spherical hole and in particular $G(\sigma) = g(\sigma)$ at $r = \sigma$. Now we will relate $P_0(r)$ to the reversible work $W(r)$ necessary to create a hole of radius r in the liquid:

$$P_0(r) = e^{-W(r)/kT} \tag{1.29}$$

This yields

$$G(r) = \frac{1}{\rho_0 kT} \left[P + \frac{2\gamma\sigma}{r} \right] \tag{1.30}$$

where γ is the surface tension. For not too small values of r

$$G(r) = \frac{1}{\rho_0 kT} \left[P + \frac{2\gamma_0}{r} + \frac{4\gamma_0\,\delta\sigma}{r^2} \right] \tag{1.31}$$

since γ is given as $\gamma = \gamma_0[1 + (2\delta\sigma/r)]$ thermodynamically, where γ_0 and δ are parameters to be determined. For $r < \sigma/2$ it is easily shown that

$$G(r) = \frac{1}{1 - \frac{4}{3}\pi r^3 \rho_0} \tag{1.32}$$

Subject to the approximations made in the theory, Eq. (1.31) is valid down to $r = \sigma/2$. Using the fact that $G(r)$ and its first derivatives are continuous at $r = \sigma/2$, then γ_0, δ, and therefore $G(\sigma)$ can be evaluated. The resulting equation of state is

$$\frac{P}{\rho_0 kT} = \frac{1 + x + x^2}{(1 - x)^3} \quad \text{where} \quad x = \left(\frac{\pi}{6}\right)(\rho_0 \sigma^3) \tag{1.33}$$

This equation is in remarkably good agreement with the machine calculations of Wood and Jacobson [18], and Wainwright and Alder [19]. It is interesting to note that Eq. (1.33) is identical to the equation of state obtained from the generalized P.Y. equation [20].

The scale particle theory has been successfully applied to several real fluids [17] and agreement with experiment is good. Although there is insufficient proof as to why this simple theory with its parameters works, still it has been the most widely applied of all the formal theories of liquid.

REFERENCES

[1] M. Born and H. S. Green, *Proc. Roy. Soc.*, **A188**, 10 (1946).
[2] J. Yvon, *Actualities Scientifiques et Industrielles*, Hermann et Cie, Paris, 1935.
[3] J. G. Kirkwood, *J. Chem. Phys.*, **3**, 300 (1935).

[4] N. H. Bogoliubov, in Part A of *Studies in Statistical Mechanics* Vol. 1, edited by
       J. de Boer and G. E. Uhlenbeck, North-Holland Publishing Co., Amsterdam, 1962.
[5] R. W. Hart, R. Wallis, and L. Pode, *J. Chem. Phys.*, **19**, 139 (1951); G. R. Rushbrooke
       and H. I. Scoins, *Phil. Mag.*, **42**, 582 (1951); B. R. A. Nijboer and L. van Hove,
       Phys. Rev., **85**, 777 (1952).
[6] A. E. Rodriguez, *Proc. Roy. Soc. (London)*, **A239**, 373 (1957); G. H. A. Cole, *J. Chem.
       Phys.*, **34**, 2016 (1961).
[7] J. G. Kirkwood, E. K. Maun, and B. J. Alder, *J. Chem. Phys.*, **18**, 1040 (1950).
       J. G. Kirkwood, V. A. Lewinson, and B. J. Alder, *J. Chem. Phys.*, **20**, 929 (1952).
       R. W. Zwanzig, J. G. Kirkwood, K. F. Stripp, and I. Oppenheim, *J. Chem. Phys.*,
       21, 1268 (1953); *ibid.*, **22**, 1625 (1954).
[8] E. Meeron, *Phys. Fluids*, **1**, 139 (1958).
       E. Meeron and E. R. Rodemich, *ibid.*, **1**, 246 (1958).
       E. Meeron, *J. Math. Phys.*, **1**, 192 (1960); *Physica*, **26**, 445 (1960).
[9] J. M. J. Van Leeuwen, J. Groenweld, and J. de Boer, *Physica*, **25**, 192 (1959).
[10] T. Morita and K. Hiroike, *Progr. Theoret. Phys. (Kyoto)*, **23**, 1003 (1960).
       K. Hiroike, *ibid.*, **24**, 317 (1960).
       T. Morita and K. Hiroike, *ibid.*, **25**, 537 (1961).
[11] J. K. Percus and G. J. Yevick, *Phys. Rev.*, **110**, 1 (1958).
[12] D. Pines and D. Bohm, *Phys. Rev.*, **85**, 338 (1951).
[13] A. A. Broyles, *J. Chem. Phys.*, **33**, 456 (1960);
                                         ibid., **34**, 359, 1068 (1961)
                                         ibid., **35**, 493 (1961).
       A. A. Broyles, S. U. Chung, and H. L. Sahlin, *J. Chem. Phys.*, **37**, 2462 (1962).
[14] M. S. Wertheim, *Phys. Rev. Letters*, **10**, 321 (1963).
[15] E. Thiele, *J. Chem. Phys.*, **39**, 474 (1963).
[16] H. Reiss, H. L. Frisch, and J. L. Lebowitz, *J. Chem. Phys.*, **31**, 369 (1959); H. Reiss,
       H. L. Frisch, E. Helfand, and J. L. Lebowitz, *ibid.*, **32**, 119 (1960); E. Helfand,
       H. Reiss, H. L. Frisch, and J. L. Lebowitz, *ibid.*, **33**, 1379 (1960); E. Helfand, H. L.
       Frisch, and J. L. Lebowitz, *ibid.*, **34**, 1037 (1961).
[17] F. H. Stillinger, Jr., *J. Chem. Phys.*, **35**, 1581 (1961).
       H. Reiss and S. W. Mayer, *ibid.*, **34**, 2001 (1961).
       S. W. Mayer, *ibid.*, **35**, 1513 (1961); **38**, 1803 (1963).
       S. J. Yosim and B. B. Owens, *J. Chem. Phys.*, **39**, 2222 (1963).
[18] W. W. Wood and J. D. Jacobson, *J. Chem. Phys.*, **27**, 1207 (1957).
[19] T. E. Wainwright and B. J. Alder, *U.S. Atomic Energy Commission Report Con-
       tract No. W-7405-eng-48*, Lawrence Radiation Laboratory, Livermore, University of
       California.
[20] J. L. Lebowitz, *Phys. Rev.*, **133**, A895 (1964).
       J. L. Lebowitz, *Bull. Am. Phy. Soc.*, **8**, 329 (1963).

chapter 2

SOME MODELS OF
THE LIQUID STATE

2.1 INTRODUCTION

This chapter is limited to a description of models of liquids used in making statistical mechanical calculations. Accordingly, Bernal's [1] geometrical or spatial model is not considered here. We will be chiefly concerned with early versions of the cell theory, the hole theory, the tunnel theory, and their refinements. The model approach is frequently called the "partition function approach," that is, to carry out the method of computing the properties of the system, the model should lead to a mathematical expression, the partition function, f.

The partition function is related to the Helmholtz free energy by the equation

$$A = -kT \ln f_N$$

where

$$f_N = \frac{1}{h^{3N}N!} \int \exp \frac{-H}{kT} \, dq_N \, dp_N \qquad (2.1)$$

Here H is the sum of the kinetic and potential energies of the N molecules expressed in terms of their coordinates q_N and their moments p_N, that is,

$$H = \sum_{i=1}^{N} \frac{p_i^2}{2m} + U \qquad (2.2)$$

where U is the total potential energy, which depends on the coordinates. To obtain the partition function, f_N, for a real fluid, we first integrate Eq. (2.1) over the momenta, p_N, which gives

$$f_N = \left(\frac{2\pi m kT}{h^2}\right)^{(3/2)N} \frac{1}{N!} \int \exp\left(-\frac{U}{kT}\right) dq_N \qquad (2.3)$$

The Helmholtz free energy, $A = -kT \ln f_N$, can be written as the sum of two parts:

$$A = A(q_N) + A(p_N) \tag{2.4}$$

where

$$A(p_N) = -\tfrac{3}{2} NkT \ln \left(\frac{2\pi mkT}{h^2} \right) \tag{2.4a}$$

and

$$A(q_N) = -kT \ln Q_N \quad \text{with} \quad Q_N = \frac{1}{N!} \int \exp \left(\frac{-U}{kT} \right) dq_N \tag{2.4b}$$

The two parts of the free energy, $A(q_N)$ and $A(p_N)$, are derived from the momenta and positional coordinates, respectively. $A(p_N)$ depends only on temperature while $A(q_N)$ is a function of both temperature and volume and is the chief source of the difficulties.

2.2 SIMPLE CELL THEORY [2]

The direct evaluation of the partition function for the liquid state is impossibly long. Thus, it is necessary to adopt a more devious strategy. Cell theories are based on an attempt to break up the configurational partition function Q_N into a product of identical similar integrals, one for each molecule. To do this, we attempt to express the potential energy of a molecule in a form which does not involve integration over the other $(N - 1)$ coordinates. We are guided by the knowledge that each molecule in a liquid or compressed gas spends much of its time confined by its neighbors in a comparatively restricted region. We picture the neighboring or wall molecules as forming a " cell " or a " wall " in which the central molecule, the wanderer, moves. According to this view, the local molecular environment in a liquid is not very different from that in a solid, although there is no long-range order and the liquid molecule has only slightly more freedom than the solid. We shall further assume that all the cells are identical, and that each contains just one molecule. For rigid spherical molecules, the configurational partition function is given simply by Eq. (2.5):

$$Q_N = \left(\frac{1}{N!} \right) \sum\nolimits^* v_f^N \tag{2.5}$$

Here the summation $\sum^*$ is to be taken over all arrangements of the N molecules in the N cells and the "free volume" v_f is the volume available to the center of a molecule in its cell. Since there are $N!$ different arrangements of N molecules in N cells, Eq. (2.5) gives

$$Q_N = v_f^N \tag{2.6}$$

To evaluate the free volume, let us assume the following: The free volume v_f has a complicated shape which is determined by the fact that the center of the wanderer molecule cannot come closer than σ to the center of a wall molecule. To simplify still further, we assume that the free volume can be approximated by a sphere of radius $(a - \sigma)$, that is,

$$v_f = \tfrac{4}{3}\pi(a - \sigma)^3 \tag{2.7}$$

Here a is the distance between the centers of neighboring molecules. Now, to express Q_N in terms of the volume of the liquid, we assume close-packed structures, thus obtaining

$$A(q_N) = -NkT \ln v_f = -NkT \ln \left\{\frac{4}{3}\pi(a - \sigma)^3\right\}$$

$$= -NkT \ln \frac{4}{3}\pi\left\{\left(\frac{\sqrt{2}\,V}{N}\right)^{1/3} - \left(\frac{\sqrt{2}\,V_s}{N}\right)^{1/3}\right\}^3 \tag{2.8}$$

Here V_s and V are the molar volumes of the solid at $0°K$ and of the liquid, respectively.

Let us calculate the equation of state for a liquid. Using the relation $P = -(\partial A(q_N)/\partial V)_T$, we obtain

$$\frac{PV}{NkT} = \frac{1}{\left\{1 - \left(\dfrac{V_s}{V}\right)^{1/3}\right\}} \tag{2.9}$$

The fact that the pressure goes to infinity when V approaches V_s and to the ideal gas value when V is very large indicates the qualitative validity of Eq. (2.9). However, for liquids at ordinary pressures, it fails hopelessly. For example, at the triple point, the value of PV/RT would be approximately unity if V were the molar volume of the vapor. As V is actually the molar volume of the liquid, PV/RT is much less than unity, but Eq. (2.9) yields values much greater than unity. This failure arises from the many simplifications and approximations which were made. One of the serious simplifications is the highly artificial nature of the hard-sphere potential function which neglects the effect of the attractive forces.

Clearly, our model will be improved by the Lennard-Jones and Devonshire type [3] of cell theory which takes account of this attractive potential. However, a much simpler approach was suggested by Eyring and Hirschfelder [4]. They conclude that U for many liquids can be represented quite accurately by a formula, $U = -a(T)/V$ where $a(T)$ is a function of temperature only. Making this assumption, we obtain

$$f(q_N) = -NkT \ln v_f - \frac{a(T)}{V} \tag{2.10}$$

which leads to what Hirschfelder has called the Eyring Equation of State:

$$P = \frac{RT}{V(1 - (V_0/V)^{1/3})} - \frac{a(T)}{V^2}$$

or

$$\left(P + \frac{a(T)}{V^2}\right)(V - V_0^{1/3} \cdot V^{2/3}) = RT \qquad (2.11)$$

This equation is a great improvement over the previous one. Its form is analogous to Van der Waal's equation and provides a useful semiempirical description of the properties of a liquid.

2.3 THE CELL THEORY OF LENNARD-JONES AND DEVONSHIRE

A further improvement on the simple cell theory outlined above was based on a realistic potential function developed in two papers [3] by Lennard-Jones and Devonshire. They used the Lennard-Jones (6-12) potential and calculated the properties of liquids and dense gases.

The Lennard-Jones theory assumes that a molecule in a liquid spends much of its time in the cell formed by its neighbors at their equilibrium positions. In order to simplify the calculation of the free volume, they further assumed that the Z nearest neighbors can be treated as if they were uniformly smeared over a spherical surface of radius equal to the nearest neighboring distance, a. Then, the partition function (for a system of N molecules) becomes

$$f_N = \frac{(2\pi mkT)^{(3/2)N}}{h^{3N}} \exp\left[-\frac{N\psi(0)}{2kT}\right] v_f^N [J(T)]^N \qquad (2.12)$$

where $J(T)$ is the partition function for the internal degrees of freedom of the molecule, and v_f is the free volume which is given by

$$v_f = \int_{\text{cell}} \exp\left[-\frac{(\psi(r) - \psi(0))}{kT}\right] dr \qquad (2.13)$$

Here, ψ is a pair interaction potential. A simple spherical molecule, such as argon, interacts with an identical molecule according to the Lennard-Jones (6-12) potential, i.e.,

$$\psi(r) = 4\varepsilon\left[\left(\frac{\sigma}{r}\right)^{12} - \left(\frac{\sigma}{r}\right)^6\right] \qquad (2.14)$$

where ε and σ are energy and distance parameters and are tabulated by Hirschfelder *et al.* [5]. This (6-12) potential was used in their calculations. In accord with the above assumptions, the field $\psi(r)$ in which a molecule moves

in its cell is the field of the surrounding molecules averaged over all directions. The cell geometry is indicated in Figure 2-1. The area of the ring is

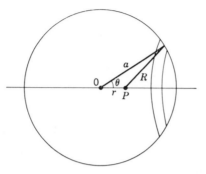

Figure 2-1   Cell geometry in the Lennard-Jones and Devonshire model.

$2\pi a^2 \sin \theta \, d\theta$ and therefore the number of smeared nearest neighbors in the
ring is $(Z/2) \sin \theta \, d\theta$. Hence,

$$\psi(r) = \int_0^\pi 4\varepsilon \left\{ \left(\frac{\sigma}{R} \right)^{12} - \left(\frac{\sigma}{R} \right)^6 \right\} \frac{Z}{2} \sin \theta \, d\theta \qquad (2.15)$$

where

$$R^2 = r^2 + a^2 - 2ar \cos \theta$$

For r fixed, we have

$$2R \, dR = 2ar \sin \theta \, d\theta$$

and

$$\psi(r) = \frac{2Z\varepsilon}{ar} \int_{a-r}^{a+r} \left\{ \left(\frac{\sigma}{R} \right)^{12} - \left(\frac{\sigma}{R} \right)^6 \right\} R \, dR \qquad (2.16)$$

Integrating Eq. (2.16) and using the relation

$$\psi(\text{o}) = 4Z\varepsilon \left\{ \left(\frac{\sigma}{a} \right)^{12} - \left(\frac{\sigma}{a} \right)^6 \right\} \qquad (2.17)$$

we obtain

$$\psi(r) - \psi(\text{o}) = Z\varepsilon\sigma^{12} \left\{ \frac{1}{5ar} \left[\frac{1}{(a-r)^{10}} - \frac{1}{(a+r)^{10}} \right] - \frac{4}{a^{12}} \right\}$$

$$+ Z\varepsilon\sigma^6 \left\{ \frac{1}{2ar} \left[\frac{1}{(a+r)^4} - \frac{1}{(a-r)^4} \right] + \frac{4}{a^6} \right\} \qquad (2.18)$$

When hexagonal close packing is assumed, then $Z = 12$ and

$$V/N = a^3/\sqrt{2}. \tag{2.19}$$

If we choose the volume of the cell so that it is equal to V/N, the free volume v_f in Eq. (2.12) becomes

$$v_f = 2\sqrt{2}\left(\frac{V}{N}\right)g \tag{2.20}$$

and

$$g = \int_0^{0.30544} \sqrt{y} \exp\left\{-\frac{[\psi(y) - \psi(a)]}{kT}\right\} dy \tag{2.21}$$

where $y = (r/a)^2 = (r^*/a^*)^2$, $r^* = r/\sigma$, and $a^* = a/\sigma$. Under these conditions, Eq. (2.18) can be written

$$\psi(r) - \psi(o) = 12\varepsilon\left\{\frac{l(y)}{v^{*4}} - \frac{2m(y)}{v^{*2}}\right\} \tag{2.22}$$

From the relations Eqs. (2.21) and (2.22), g becomes

$$g = \int_0^{0.30544} \sqrt{y} \exp\left\{-\frac{12}{T^*}\left[\frac{l(y)}{v^{*4}} - \frac{2m(y)}{v^{*2}}\right]\right\} dy \tag{2.23}$$

where

$$l(y) = (1 + 12y + 25.2y^2 + 12y^3 + y^4)(1 - y)^{-10} - 1$$

$$m(y) = (1 + y)(1 - y)^{-4} - 1$$

$$v^* = \frac{v}{\sigma^3} = \frac{v}{N\sigma^3} \quad \text{and} \quad T^* = kT/\varepsilon \tag{2.24}$$

Then, using Eq. (2.12) the pressure equation becomes

$$\frac{PV}{NkT} = -\frac{v}{2kT}\left(\frac{\partial\psi(o)}{\partial v}\right)_T + v\left(\frac{\partial \ln v_f}{\partial v}\right)_T \tag{2.25}$$

Here $v = a^3/\sqrt{2}$. The first term is the potential part of the pressure resulting from the potential energy of all the molecules located at the center of the cell, and the second term is due to the pressures resulting from the motion of the molecules in their cells. Carrying out these calculations, we obtain

$$\frac{PV}{NkT} = 1 + \frac{24}{T^*}\left[\frac{1}{v^{*4}}\left(1 + \frac{2g_l}{g}\right) - \frac{1}{v^{*2}}\left(1 + \frac{2g_m}{g}\right)\right] \tag{2.26}$$

where g is given by Eq. (2.20) and

$$g_l(v^*, T^*) = \int_0^{0.30544} l(y)\sqrt{y}\, \exp\left\{-\frac{Z}{T^*}\left[\frac{l(y)}{v^{*4}} - \frac{2m(y)}{v^{*2}}\right]\right\} dy \quad (2.27)$$

$$g_m(v^*, T^*) = \int_0^{0.30544} m(y)\sqrt{y}\, \exp\left\{-\frac{Z}{T^*}\left[\frac{l(y)}{v^{*4}} - \frac{2m(y)}{v^{*2}}\right]\right\} dy \quad (2.28)$$

Several authors [3], [6], [7] have tabulated the above integral.

Originally, Lennard-Jones and Devonshire only considered the effect of shells beyond the first on the lattice energy, not on the free volume. So far, the most extensive calculations of the Lennard-Jones and Devonshire equation of state have been carried out by Wentorf et al. [8]. They included the effect of the three neighboring shells on both the lattice energy and on the free volume and obtained the following equation of state

$$\frac{PV}{NkT} = 1 - \frac{12}{T^*}\left[\frac{2.4090}{v^{*2}} - \frac{2.0219}{v^{*4}}\right] - \frac{48}{T^*}\left[\frac{1}{v^{*2}}\frac{G_M}{G} - \frac{1}{v^{*4}}\frac{G_L}{G}\right] \quad (2.29)$$

and G, G_L, and G_M are integrals like g, g_l, and g_m except that $l(y)$ and $m(y)$ are replaced by the functions

$$L(y) = l(y) + \frac{1}{128}\, l\left(\frac{1}{2}\, y\right) + \frac{2}{729}\, l\left(\frac{1}{3}\, y\right) \quad (2.30)$$

$$M(y) = m(y) + \frac{1}{16}\, m\left(\frac{1}{2}\, y\right) + \frac{2}{27}\, m\left(\frac{1}{3}\, y\right) \quad (2.31)$$

The excess of a property is defined as the difference between the real property and the ideal gas property. Hence, the excess reduced energy $E^E/N\varepsilon$ and the excess reduced entropy S^E/Nk are derived as follows:

$$S^E/Nk = \ln\left[2\sqrt{2}\,\pi\,\frac{G}{e}\right] + \frac{12}{T^*}\left[\frac{1}{v^{*4}}\frac{G_L}{G} - 2\frac{1}{v^{*2}}\frac{G_M}{G}\right] \quad (2.32)$$

and

$$\frac{E^E}{NkT} = \frac{6}{T^*}\left[\frac{1.0109}{v^{*4}} - \frac{2.4090}{v^{*2}}\right] + \frac{12}{T^*}\left[\frac{1}{v^{*4}}\frac{G_L}{G} - 2\frac{1}{v^{*2}}\frac{G_M}{G}\right] \quad (2.33)$$

The above theory also can be applied to quantum liquids. The only modification is that the integrals in Eq. (2.1) must be replaced by a sum and the energy levels must be calculated by solving Schrödinger's equation. Levelt and Hurst [9] have done this for liquid hydrogen for the single volume of $v^* = \frac{5}{3}$.

2.4 HOLE OR FREE VOLUME THEORY

Experiments show that the density and the apparent number of nearest neighbors ordinarily decrease upon melting. Further, as the temperature is raised, both density and the apparent number of nearest neighbors decrease. Also, the liquid has an entropy increase upon melting. The above facts suggest that allowance be made for the presence of vacant lattice sites or holes if we use a lattice picture of the structure of the liquid. An alternative explanation invokes the presence of a substantial amount of communal entropy in the liquid [10] which is not present in the solid but is present in the gas. The concept of the presence of holes and their more or less random distribution explains at least part of the entropy increase on melting. Furthermore, this provides a basis for understanding the ease of flow or the diffusion mechanism in terms of a "vacancy diffusion" which originally led Eyring [11] to postulate the presence of holes in liquids. The additional disorder and entropy are supplemented by "communal entropy" in the cell theory.

The concept of Cernuschi and Eyring [11] of vacancies in a liquid quasi-lattice has been reviewed by Rowlinson and Curtiss [12]. The volume V of the liquid is divided into $L > N$ cells so that $v = V/N$ is the volume per molecule and $\omega = V/L$ is the volume per cell. Further, we suppose that the cells are large enough so that only nearest neighbor interaction need be considered. Because of the holes in the liquid, the coordination number of a particle i will be decreased, that is,

$$Z_i = y_i Z$$

where y_i is some number between zero and unity.

Suppose that the free volume of the molecule i only depends on the value of y_i, then Eq. (2.12) becomes

$$f_N = \left[\frac{L!}{N!(L-N)!} \right] \frac{(2\pi m kT)^{(3/2)N}}{h^{3N}} \exp\left[\frac{-yNZ\phi(\mathrm{o})}{2kT} \right] (v_f)^N$$

$$\text{here}\quad \psi(\mathrm{o}) = Z\phi(\mathrm{o}) \quad (2.34)$$

Equation (2.34) can be evaluated by the Bragg-Williams or quasi-chemical approximation. For the evaluation of the partition function, we have to know the dependence of v_f on y. The Lennard-Jones and Devonshire values for the free volume are the only widely tabulated values and have been exclusively used so far. If all the cells around the molecule are occupied, then $y = 1$ and v_f is identical with v_f^0, as tabulated by Wentorf et al. [8]. If all the surrounding cells are empty, then $y = \mathrm{o}$, and v_f becomes the volume of the cell, q.

Actually, v_f does not depend on y in a simple way. However, the following linear dependency of v_f has been assumed by various authors.

(a) Cernuschi and Eyring [11]

$$v = v_f^0 \qquad (2.35a)$$

(b) Ono [13]

$$\ln v_f = y \ln v_f^0 + (1 - y) \ln q \qquad (2.35b)$$

(c) Peek and Hill [14]

$$v_f = v_f(\bar{y}) \qquad (2.35c)$$

where $\bar{y}$ is the average nonempty site around any molecule.

(d) Rowlinson and Curtiss [12]

$$\ln v_f = \ln v(\bar{y}) + (y - \bar{y})\left[\frac{\partial}{\partial y} \ln v_f\right]_{y=\bar{y}} \qquad (2.35d)$$

(e) Henderson [15]

$$v_f = yv_f^0 + (1 - y)q \qquad (2.35e)$$

The approximations of Cernuschi and Eyring, and Ono are alike in that q is chosen as equal to the cell volume of the solid at the melting point, whereas in the approximations of Peek and Hill, and Rowlinson and Curtiss, the value of q is chosen so that the Helmholtz free energy is a minimum for each volume and temperature. The latter procedure is somewhat preferable theoretically, but it is found that q increases much too rapidly with increasing volume in the calculation, with the result that too small a proportion of holes is predicted. Grindley [16] has recently been able to evaluate the partition function explicitly in the quasi-chemical approximation without assuming a linear dependence of $\ln v_f$ on y. In this investigation, q was determined by minimizing the free energy. The results of all these calculations can be seen in Table 2-1. The free energy and the other results do not differ appreciably

Table 2-1 Calculated Values of the Critical Point Properties

Mean value for	T_c^*	V_c^*	P_c^*	$P_c^* V_c^* / T_c^*$	y_c
Ne, Ar, N_2	1.28	3.15	0.119	0.293	~0.5
L.J.&D.	1.30	1.77	0.434	0.591	1.000
C.&E.	2.75	2.00	0.470	0.342	0.544
Ono	0.75	2.00	0.128	0.342	0.544
P.&H.	1.18	3.25	0.261	0.825	0.825
R.&C.	—	—	—	—	—
Grindley	1.30	1.77	0.434	0.591	1.000
Henderson	1.41	3.39	0.139	0.333	0.321

from the unmodified theory of Lennard-Jones and Devonshire. Henderson's results are quite good and show a significant improvement over previous attempts. The method works well in the dense gas region, as shown in Figure 2-2, where the calculated values of the second virial coefficient are compared with the other results.

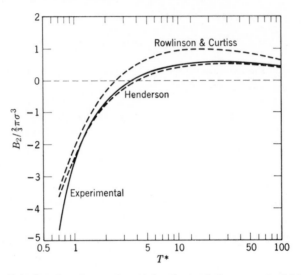

Figure 2-2 Calculated and experimental values of the second virial coefficient (after Henderson [15])

Besides these examples, a great number of other modifications of the free volume theory have been developed. However, the theories outlined are representative. Kirkwood [17] has shown that the statistical mechanical foundation of the Lennard-Jones and Devonshire theory is a first approximation to a variational treatment for the calculation of the free energy. According to recent investigations of Mayer and Careri [18], Dahler, Hirschfelder, and Thacher [19], and Barker [20], this method is not satisfactory.

Other approaches to correct the defects of the Lennard-Jones and Devonshire theory directly are the nonspherialization of the cell, the correct calculation of the communal entropy, and the consideration of correlation with neighboring molecules. Buehler *et al.* [21] have carried out the numerical integration over the dodecahedral cell for nonattracting rigid sphere molecules. They find a free volume correction factor of up to 1.4 in the critical region. Pople [22a] and Janssens and Prigogine [22b] have made the communal entropy correction by considering the probability of finding two molecules in one cell. However, Pople found that even in the dense gas region, double occupancy makes only a small contribution to the free energy. Weissmann and Mazo [23] examined double occupancy effects in much more detail than did Pople and Janssens and Prigogine and their results give corrections in the right direction in the range of liquid densities. However, for the entropy, the corrections are too small.

These facts suggest the need of further refinements such as multiple occupancy and of molecular correlation terms. It is interesting that the inclusion of the correlation term by Barker [20] leads to a solid-liquid transition, since no other modification of the free volume theory succeeds in this respect. Similar considerations were investigated by de Boer, Cohen, and colleagues [24]. Recently, Chung and Dahler [25], [26] developed the worm theory of the liquid state assuming that the molecular interactions were pair additive and were given by the L.J. (6-12) potential. The initial calculations were restricted to high density regions. More recently, Chung and Dahler [27] extended the calculation over a broad range of physical conditions. Although the theory does represent a considerable improvement over the cell theory, it provides good estimates only for a restricted range at very high pressures. This shows that there is a higher degree of molecular ordering in the worm-model representation than that characteristic of a real liquid.

2.5 TUNNEL THEORY

One can imagine an irregular lattice model replacing the regular lattice but still having sufficient regularity or simplicity to permit a simple description, yet being disordered in a more fundamental way than the cell or hole models permit. One such model is the tunnel theory proposed by Barker [28] which is essentially a one-dimensional disordered cell theory. The basic idea of the tunnel theory is that the whole system of molecules is imagined to be divided into subsystems consisting of lines of molecules moving almost one-dimensionally in tunnels whose walls are formed by neighboring lines. It is assumed that the longitudinal and transverse motions of the molecules in a given tunnel may be treated as separable. Then, the partition function is the product of the partition functions for longitudinal and transverse motions in the tunnel. The partition function for the one-dimensional longitudinal motion can be expressed as

$$f_M = \left(\frac{(2\pi m k T)^{1/2}}{h} \right)^M \left[\frac{1}{M!} \int_0^{lM} \cdots \int_0^{lM} \exp \left[-\frac{U(x)}{kT} \right] dx_1 \cdots dx_M \right] \quad (2.36)$$

where $U(x)$ is the potential energy of the one-dimensional system of M molecules in the length lM. When there are K such tunnels in a system, the total partition function will be f_M^K. The partition function for the transverse motion can be evaluated in terms of a "free area," a_f, exactly analogous to the "free volume" in the cell model. Then, the total partition function is

$$f_N = \frac{(2\pi mkT)^{(3/2)N}}{h^{3N}} \left[a_f \cdot \exp\left(-\frac{\psi(o)}{2kT}\right) \right]^N \cdot f_M^K [J(T)]^N \qquad (2.37)$$

where

$$a_f = \int \exp\left\{-(\psi(r) - \psi(o))/kT\right\} dr \qquad (2.38)$$

In the above equation, $\psi(r)$ is the potential energy of interaction of a given molecule at a distance r from the axis of its tunnel, and this is summed over all molecules except those in the tunnel. The one-dimensional partition function defined in Equation (2.36) is written as:

$$-\frac{1}{N} \ln\left(f_M\left(\frac{(2\pi mkT)^{1/2}}{h}\right)^{-M}\right) = \frac{A_1^*}{RT} + \frac{4}{T^*}\left[0.0003\left(\frac{\sigma}{l}\right)^{12} - 0.0173\left(\frac{\sigma}{l}\right)^6\right]$$

$$(2.39)$$

The second term on the right is the Helmholtz free energy due to the non-nearest neighbor interaction; the first term, that due to nearest neighbor interactions, is

$$A_1^* = -\frac{P_1 l}{kT} - \ln\left[\int_0^\infty \exp\left\{-[U(x) + P_1 x]/kT\right\} dx\right] \qquad (2.40)$$

Here P_1 is the one-dimensional pressure which is related to the intermolecular distance l by Eq. (2.41)

$$l = \frac{\displaystyle\int_0^\infty x \exp\left\{-[U(x) + P_1 x]/kT\right\} dx}{\displaystyle\int_0^\infty \exp\left\{-[U(x) + P_1 x]/kT\right\} dx} \qquad (2.41)$$

The values of $\psi(o)$ and a_f in Eq. (2.37) can be obtained assuming the hexagonal close-packed lattice with the lattice distance a as follows:

$$V/N = (\sqrt{3}/2)a^2 l \qquad (2.42)$$

$$\psi(o)/kT = 4.652C - 7.925B \qquad (2.43)$$

here

$$B = \frac{4\varepsilon a}{lkT}\left(\frac{\sigma}{a}\right)^6, \qquad C = \frac{4\varepsilon a}{lkT}\left(\frac{\sigma}{a}\right)^{12} \qquad (2.44)$$

and a_f is given by

$$a_f = a^2 S \qquad (2.45)$$

where

$$S = 2\pi \int_0^{0.52} \exp - \left[\frac{\psi(y) - \psi(o)}{kT}\right] y \, dy \quad \text{here} \quad y = \frac{r}{a} \qquad (2.46)$$

Using the relation $A = -kT \ln f_N$ and Eqs. (2.39) through (2.46), we obtain the following equation for the pressure:

$$\frac{PV}{NkT} = \frac{1}{T^*}\left(\frac{V_0}{V}\right)^4\left(20.938 + 9\frac{S_C}{S}\right)$$

$$- \frac{1}{T^*}\left(\frac{V_0}{V}\right)^2\left(23.880 + 6\frac{S_B}{S}\right) + \frac{0.3497}{W}\left(\frac{V_0}{V}\right)^{-1/3} + \frac{2}{3} \qquad (2.47)$$

In this equation, the relation $l = a = \sigma(2V/\sqrt{3}\,V_0)^{1/3}$ has been introduced, where

$$S_B = \frac{\partial S}{\partial B}, \quad S_C = -\frac{\partial S}{\partial C}, \quad W = \frac{kT}{P_1\sigma}, \quad \text{and} \quad V_0 = N\sigma^3$$

S, S_B, S_C, and W have been tabulated by Barker [28b]. The zero-pressure equilibrium volume is found by setting P equal to zero and solving Eq. (2.47) for (V/V_0).

In Table 2-2, some of the calculated results for reduced volume, reduced

Table 2-2 Theoretical and Experimental Properties at $kT/\varepsilon = 0.70$

	Reduced volume (V/V_0)	Reduced excess energy $(E^E/N\varepsilon)$	Reduced excess entropy (S^E/Nk)
Tunnel theory	1.184	−5.9	−4.8
Liquid argon	1.181	−5.96	−3.64
L.J.&D. theory	1.037	−7.32	−5.51
Solid argon	1.035	−7.14	−5.33

excess energy, and reduced excess entropy for zero pressure and $kT/\varepsilon = 0.70$ are compared with experiment. The volume and energy calculated from the tunnel theory are close to the experimental values for liquid argon, thus showing that the tunnel theory is a theory of a liquid rather than a solid. However, the value of the entropy gives poor agreement. This may be because the tunnel theory underestimates the volume of configuration space associated with these structures, that is, a difficulty lies in the assumption that the molecules in different tunnels move independently. Further progress with the tunnel theory will require the treatment of correlation effect analogous to that of the cell theory. Recently, this theory has been applied to quantum liquids by Oden and Henderson [29] assuming the hard sphere system. The thermodynamic properties were calculated and the conclusion made that quantum corrections are still present even at relatively high temperatures.

Finally, we only list several references [28a, 30, 31] to the calculations which permit the direct comparisons between these models discussed in this chapter and those of the radial distribution function.

REFERENCES

[1] J. D. Bernal, *Nature*, **183**, 141 (1959).
        ibid., **185**, 68 (1960).
        Sci. Am., **203**, 124 (August, 1960).

[2] R. J. Buehler, R. H. Wentorf, J. O. Hirschfelder, and C. F. Curtiss, *J. Chem. Phys.*, **19**, 61 (1951).

[3] J. E. Lennard-Jones and A. F. Devonshire, *Proc. Roy. Soc.*, **163**, 53 (1937).
        ibid., **165**, 1 (1938).

[4] H. Eyring and J. O. Hirschfelder, *J. Phys. Chem.*, **41**, 249 (1937).

[5] J. O. Hirschfelder, C. F. Curtiss, and R. B. Bird, *Molecular Theory of Gases and Liquids*, Wiley, New York, 1964, pp. 1110–1113 and 1212–1215.

[6] I. Prigogine and S. Raulier, *Physica*, **9**, 396 (1942); I. Prigogine and G. Garikian, *J. Chim. Phys.*, **45**, 273 (1948).

[7] T. L. Hill, *J. Phys. and Colloid Chem.*, **51**, 1219 (1947).

[8] R. H. Wentorf, Jr., R. J. Buehler, J. O. Hirschfelder, and C. F. Curtiss, *J. Chem. Phys.*, **18**, 1484 (1950).

[9] J. M. H. Levelt and R. P. Hurst, *J. Chem. Phys.*, **32**, 96 (1960).

[10] J. O. Hirschfelder, D. Stevenson, and H. Eyring, *J. Chem. Phys.*, **5**, 896 (1937).

[11] (a) H. Eyring, *J. Chem Phys.*, **4**, 283 (1936).
        (b) F. Cernuschi and H. Eyring, *J. Chem. Phys.*, **7**, 547 (1939).

[12] J. S. Rowlinson and C. F. Curtiss, *J. Chem. Phys.*, **19**, 1519 (1951).

[13] S. Ono, *Mem. Fac. Eng. Kyushu Univ.*, **16**, 190 (1947).

[14] H. M. Peek and T. L. Hill, *J. Chem. Phys.*, **18**, 1252 (1950).

[15] D. Henderson, *J. Chem. Phys.*, **37**, 631 (1962).

[16] J. Grindley, *Proc. Phys. Soc. (London)*, **77**, 1001 (1961).

[17] J. Kirkwood, *J. Chem. Phys.*, **18**, 380 (1950).

[18] J. E. Mayer and G. Careri, *J. Chem. Phys.*, **25**, 249 (1956).

[19] J. S. Dahler, J. O. Hirschfelder, and H. C. Thacher, *J. Chem. Phys.*, **25**, 249 (1956).

[20] J. A. Barker, *Proc. Roy. Soc. (London)*, **A230**, 390 (1955); **A237**, 63 (1956); **A240**, 265 (1957); **A241**, 547 (1958).

[21] R. J. Buehler, R. H. Wentorf, Jr., J. O. Hirschfelder, and C. F. Curtiss, *J. Chem. Phys.*, **19**, 61 (1951).

[22] (a) J. A. Pople, *Phil. Mag.*, **42**, 459 (1951).
        (b) P. Janssens and I. Prigogine, *Physica*, **16**, 895 (1950).

[23] M. Weissmann and R. M. Mazo, *J. Chem. Phys.*, **37**, 2930 (1962).

[24]   J. de Boer, *Physica*, **20**, 655 (1954); E. G. D. Cohen, J. de Boer, and Z. W. Salsburg, *ibid.*, **21**, 137 (1955); **23**, 389 (1957); Z. W. Salsburg, E. G. D. Cohen, B. C. Rethmeier, and J. de Boer, *ibid.*, **23**, 407 (1957); E. G. D. Cohen and B. C. Rethmeier, *ibid.*, **24**, 959 (1958); J. S. Dahler and E. G. D. Cohen, *ibid.*, **26**, 81 (1966).

[25]   H. S. Chung and J. S. Dahler, *J. Chem. Phys.*, **40**, 2868 (1964).

[26]   H. S. Chung and J. S. Dahler, *J. Chem. Phys.*, **40**, 2374 (1965).

[27]   H. S. Chung and J. S. Dahler, *J. Chem. Phys.*, **43**, 2606 (1965).

[28]   (a) J. A. Barker in *The International Encyclopedia of Physical Chemistry and Chemical Physics: Topics 10, The Fluid State*, J. S. Rowlinson, ed., Vol. 1, Lattice Theories of the Liquid State, Macmillan, New York, 1963.
       (b) J. A. Barker, *Proc. Roy. Soc. (London)*, **A259**, 442 (1961).

[29]   L. Oden and D. Henderson, *J. Chem. Phys.*, **41**, 3487 (1964).

[30]   J. Corner and J. E. Lennard-Jones, *Proc. Roy. Soc.*, **A178**, 401 (1941). G. S. Rushbrooke, *Proc. Roy. Soc. Edin.*, **A60**, 182 (1940). J. S. Dahler, *J. Chem. Phys.*, **29**, 1082 (1958).

[31]   H. S. Chung and J. S. Dahler, *J. Chem. Phys.*, **43**, 2610 (1965).

chapter 3

SIGNIFICANT STRUCTURE THEORY OF LIQUIDS

3.1 INTRODUCTION

In this chapter, we discuss Eyring's extended version of his early hole theory as the significant structure approach [1]. A general review and a record of progress have been given in earlier publications [2–5].

3.2 EXPERIMENTAL BASIS OF THE MODEL

(a) Excess Volume, Fluidized Vacancies

When an ordinary liquid such as argon melts, its volume expands about 12%. In exceptional cases, such as ice, the volume decreases on melting due to a structural shrinkage which accompanies and exceeds the normal expansion. However, in spite of the increase in volume and the decrease of the coordination number upon melting, X-ray diffraction experiments [6] show no appreciable change in the nearest neighbor distance (3.8 Å for argon). Also, the X-ray data indicate that the nearest neighbors of a molecule are arranged in an orderly manner, whereas the second and third nearest neighbors are more randomly distributed, and beyond the third chaos reigns. In other words, the liquid possesses short range order but shows a total lack of long range order. This can be understood if some of the molecules are replaced by "fluidized" vacancies. No doubt some holes will be somewhat smaller than vacancies and some larger, but presumably holes of molecular size will be strongly favored because smaller holes will not provide easy access to entering molecules and larger holes will be unnecessarily wasteful of energy. An expansion of 12% corresponds to removing about every eighth molecule of liquid argon. This has two effects. First, it increases the volume simply by increasing the number of sites, at the same time keeping the intermolecular distance constant. Second, when two or more molecules share a vacancy, the lattice structure is destroyed, so that we no longer have long range order. Long range order is also blurred by a distribution of hole volumes

28

around the vacancy as a mean. Thus, a liquid has an excess volume, $V - V_s$, where V and V_s are the molar volume of the liquid and of the solid, respectively.

It should be noted that liquids have many fluidized vacancies resulting in soft space in which there is a potential hole of molecular size, while solids have a limited number of static vacancies. The following considerations support this view.

Dynamic vacancies of molecular size are likely to occur also since they leave all but nearest neighbors of the vacancy relatively undisturbed. In order to take possession of a vacancy, a nearest neighbor must have sufficient kinetic energy to push back the other neighbors competing for the vacancy, that is, it must have enough excess kinetic energy to equal or exceed the mean kinetic energy which the competing neighbors can put in the vacancy, since kinetic energy density is proportional to the pressure. The molecules that have been pushed back by the energized nearest neighbor become part of the hard volume of the liquid. The above picture is in accord with light scattering data which show that the extra volume upon melting is not present as static vacancies of the kind assumed to exist at much lower concentration in solids. The delayed nucleation of crystals suggests lack of large perfect crystalline regions in liquids, while delayed formation of vapor bubbles exhibited by bumping is evidence of a lack of vacancies large enough to nucleate bubbles.

We now consider the further characterization of fluidized vacancies. According to the law of rectilinear diameters, the mean density of a liquid and its equilibrium vapor is a linear function of temperature: at the melting point it equals half the density of the solid and decreases with increasing temperature to about one-third the solid density at melting at the critical temperature. Some decrease in the mean density due to the lattice expansion with increasing temperature is to be expected. This result indicates that the number of molecules in 1 cc of vapor equals the number of fluidized vacancies in 1 cc of the liquid, to a useful approximation. To vaporize a quantity of liquid requires the heat of vaporization, which is half the energy of all the bonds a molecule forms in the liquid. To remove a molecule from the interior of the liquid to the vapor, leaving a vacancy behind, requires twice the energy of vaporization, i.e., the sum of all the bonds the molecule made. Thus, a molecule taken from the interior of the liquid to the vapor can be returned to the surface of the liquid, regaining the energy of vaporization but leaving a vacancy which costs just the energy of vaporization. In the soft spaces of the liquid, a dynamic vacancy should move through the liquid about as freely as a molecule moves in the gas. Hence, we would expect the hole to associate with it about the same heat and entropy as does a vapor molecule. This gives the explanation of the law of rectilinear diameters. This does not mean that a hole has thermodynamic properties, but it acts puppet-wise to provide such properties to neighboring molecules. (See Figure 3.1.)

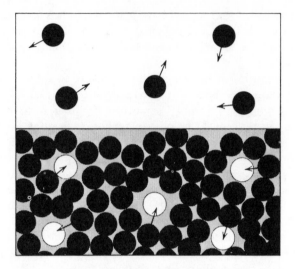

Figure 3-1 Vacancies in a liquid behave like molecules in a gas. In a liquid, vacancies move among molecules. In a gas, molecules move among vacancies. (after *International Science and Technology*, March, 1963)

(b) The Mole Fractions of Solid-Like and Gas-Like Molecules

As we have seen, a mole of liquid contains $(V - V_s)/V_s$ moles of holes of molecular size. If a vacancy is completely surrounded by molecules, we expect it will contribute gas-like properties equivalent to one gas molecule. If a vacancy is completely surrounded by vacancies, it will have no dynamic properties. To the approximation that vacancies are randomly distributed, V_s/V of the positions next to a vacancy are filled, so this fraction of the vacancies should confer gas-like properties on the liquid. Thus, the fraction

$$\left[\left(\frac{V_s}{V} \right) \left(\frac{V - V_s}{V_s} \right) \right] = \frac{V - V_s}{V}$$

of the degrees of freedom should be gas-like. The remaining fraction, V_s/V can be thought of as being solid-like. Accordingly, the heat capacity at constant volume, C_V, of a mole of argon should be given by the sum of the contribution from V_s/V moles of solid and $(V - V_s)/V$ moles of gas. Thus

$$C_V = 6 \frac{V_s}{V} + 3 \frac{V - V_s}{V} \qquad (3.1)$$

This relation was first suggested by Walter and Eyring [7]. Its validity is tested in Figure 3.2. The results are quite satisfactory, thus suggesting fluidized vacancy theory is on the right track. Other properties such as thermal conductivity, viscosity, and dielectric constants are successfully calculated in analogous fashion. These calculations will be described in later chapters. These successes led to a general development of the significant structure theory.

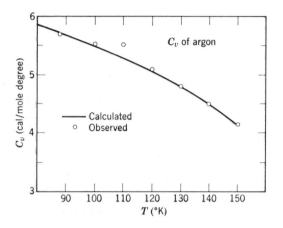

Figure 3-2 Heat capacity at constant volume for liquid argon. The solid curve represents Eq. (3.1) and the circles represent the experimental data. (after Eyring and Ree [1])

3.3 FORMULATION OF THE PARTITION FUNCTION

In view of the above, the partition function, f_N for a mole of liquid can be written as

$$f_N = (f_s)^{N(V_s/V)}(f_g)^{N\frac{V-V_s}{V}} \tag{3.2}$$

Here N is Avogardo's number; f_s and f_g are the partition functions of solid-like and gas-like degree of freedom, respectively. In deriving f_s, we have to consider a positional degeneracy factor that multiplies the usual partition function for a solid. As stated earlier, if a molecule is to have access to fluidized vacancies, it must push the competing neighboring molecules aside. When the molecule has the required energy, the additional site becomes available to it and there is a degeneracy factor equal to the number of such sites made available plus the original site. The number of additional sites will be equal to the number of vacancies around a solid-like molecule multiplied by the probability that the molecule has the required energy ε_h/N, to move into a site. Thus, the number of additional sites is

$$n_h \, e^{-\varepsilon_h/RT} \tag{3.3}$$

Here n_h is proportional to the number of vacancies, giving us

$$n_h = n(V - V_s)/V_s \tag{3.4}$$

and ε_h should be inversely proportional to the number of vacancies and directly proportional to the energy of sublimation of the solid, E_s. Thus,

$$\varepsilon_h = \frac{aE_s V_s}{V - V_s} \tag{3.5}$$

ε_h in Eq. (3.5) approaches infinity as the solid volume is approached and becomes zero for very large volumes. This is the point at which cooperative phenomena enter into significant structure theory and the resulting degeneracy factor given by Eq. (3.6), accounts for the appearance of a stable liquid phase in competition with the solid phase. The general behavior of ε_h is certainly correct but further examination of the glassy region is necessary if we are to obtain a better value for ε_h in the metastable volume range lying between the solid and liquid states. n and a are the proportionality constants which will be evaluated.

The total number of positions available to a given molecule is

$$1 + n_h \, e^{-\varepsilon_h/RT} \tag{3.6}$$

If we assume that an Einstein oscillator is an adequate representation of the lattice vibrational degree of freedom of the solid-like molecules at the temperature of interest, we can write the f_s for a monatomic liquid such as argon as follows:

$$f_s = \frac{e^{E_s/RT}}{(1 - e^{-\theta/T})^3} \cdot (1 + n_h \, e^{-\varepsilon_h/RT}) \tag{3.7}$$

where θ is the Einstein characteristic temperature, and E_s is the sublimation energy.

For the partition function of the gas-like degrees of freedom, f_g, we use the nonlocalized independent ideal gas partition function for the $(N(V - V_s)/V)$ gas-like molecules moving in the excess volume $(V - V_s)$. This gives us

$$f_g^{N\frac{V-V_s}{V}} = \left\{ \frac{(2\pi mkT)^{3/2}}{h^3} (V - V_s) \right\}^{N\frac{V-V_s}{V}} \left\{ \left(\frac{N(V - V_s)}{V} \right)! \right\}^{-1} \tag{3.8}$$

Using Eqs. (3.3) to (3.8) and substituting into (3.2), we obtain

$$f_N = \left\{ \frac{e^{E_s/RT}}{(1 - e^{-\theta/T})^3} \left(1 + n\left(\frac{V - V_s}{V_s}\right) \exp\left(- \frac{aE_s V_s}{(V - V_s)RT} \right) \right) \right\}^{N\frac{V_s}{V}}$$

$$\times \left\{ \frac{(2\pi mkT)^{3/2}}{h^3} \frac{eV}{N} \right\}^{N\frac{V-V_s}{V}} \quad (3.9)$$

Here, the last term was obtained by substituting for $((N(V - V_s))/V)!$ using Stirling's approximation $x! = (x/e)^x$.

3.4 THE THEORETICAL CALCULATION OF THE PARAMETERS n AND a AND THE APPLICATION TO INERT GASES

One of the criticisms frequently raised against this theory is that it has a number of adjustable parameters, E_s, θ, V_s, a, and n in the partition functions. For a simple liquid, the values of E_s, θ, and V_s are obtained from the solid state. The values of a and n are evaluated theoretically. The theory works without adjustable parameters for simple liquids. For polyatomic molecules involving restricted rotation and a change in the solid-like structure upon melting, a more sophisticated calculation of the parameters n and a is required or, alternatively, they can be determined from experimental data. Here, we take the monatomic liquid and show how the model leads to values for the parameters. The degeneracy factor in Eq. (3.9) involves the two parameters n and a whose values can be calculated from our model. For a simple liquid such as argon, near the melting point, the liquid approximates a solid-like lattice expanded about 12% through the addition of vacancies. Thus, near the melting point, the fraction of the neighboring positions, Z, which are empty and therefore available for occupancy is

$$Z\left(\frac{V_m - V_s}{V_m}\right) = \left(Z\frac{V_s}{V_m}\right) \cdot \frac{V_m - V_s}{V_s} = n\frac{V_m - V_s}{V_s} \quad (3.10a)$$

Hence

$$n = Z\frac{V_s}{V_m} \quad (3.10b)$$

We next calculate a. The average solid molecule has kinetic energy equal to $(\frac{3}{2})kT$. Now if a molecule is to preempt a neighboring position in addition to its original position, it must have additional kinetic energy equal to or in excess of that which the other $n - 1$ neighboring molecules would otherwise introduce into this vacancy. If an average molecule divides its time equally between two neighboring positions, this will cut its energy density in two. Since a molecule will be moving $(1/Z)$th of its time in the direction of any neighbor, the average kinetic energy of $(n - 1)$ ordinary molecules will provide a vacancy with the kinetic energy $(\frac{1}{2})(\frac{3}{2})kT(n - 1)/Z$ and this is the value which $aE_s V_s/(V_m - V_s)N$ must have at the melting point, i.e.,

$$\varepsilon_h = \frac{aE_s V_s}{V_m - V_s} = \frac{1}{2}\left(\frac{3}{2} NkT_m\right)\frac{n - 1}{Z} \qquad (3.11)$$

Since the entropy of melting per molecule of argon is very close to

$$\left(\frac{3}{2}\right)k = \frac{E_m}{NT} \qquad (3.11a)$$

and since the energy of melting per mole, E_m, is due to introducing holes into the solid with no change in kinetic energy, we have $E_m = (V_m - V_s)E_s/V_m$. Equation (3.11a) also follows if we recognize that for the liquid phase to be dynamically stable the potential energy per molecule introduced by the expan-

Table 3-1   Vapor Pressures, Molar Volumes, and Entropy of Vaporization

Temperature ($^\circ$K)	P_{calc} (atm.)	P_{obs} (atm.)	V_{calc} (cc)	V_{obs} (cc)	ΔS_{calc} (e.u.)	ΔS_{obs} (e.u.)
Argon						
83.96 (m.p)	0.6874	0.6739	28.84	28.03	20.07	19.43
87.49 (b.p)	1.040	1.000	29.36	28.69	18.90	18.65
97.76	2.883	2.682	31.04	30.15	15.92	—
Krypton						
116.0 (m.p)	0.7605	0.7220	34.31	34.13	20.14	—
119.93 (b.p)	1.0660	1.000	34.90	—	19.15	17.99
Xenon						
161.3 (m.p)	0.8372	0.804	42.24	42.68	20.25	—
165.1 (b.p)	1.0623	1.000	42.84	—	19.50	18 29

E_s, θ, and V_s values used

	Ar	Kr	Xe
E_s (cal/mole)	1888.6	2740	3897.7
V_s (cc/mole)	24.98	29.6	36.5
θ ($^\circ$K)	60.0	45.0	39.2

sion in melting must be balanced by the kinetic energy per molecule. Thermo-dynamic stability is something quite different and has to do with the relative values of the Gibbs free energy of the solid and liquid phases. Accordingly, we can write:

$$\frac{aE_s V_s}{V_m - V_s} = \frac{1}{2} E_m \frac{n-1}{Z} = \frac{1}{2} \frac{V_m - V_s}{V_m} E_s \frac{n-1}{Z} \tag{3.12a}$$

Hence,

$$a = \frac{n-1}{Z} \frac{1}{2} \frac{(V_m - V_s)^2}{V_m V_s} \tag{3.12b}$$

Equations (3.10) and (3.12) give $a = 0.0054$ and $n = 10.7$ for argon. At the early stage of the significant structure approach in 1960, Fuller et al. [8] used $a = 0.00534$ and $n = 10.8$ to get the best empirical results for argon. This agreement is highly satisfactory. Some of the properties of the monatomic liquid are shown in Table 3-1. In this calculation, the values of E_s, θ, and V_s were obtained from the solid data, and Eqs. (3.12) and (3.10) were used to calculate n and a. Thus, the introduction of adjustable parameters is avoided. The results show excellent agreement between theory and experiment.

3.5 IMPROVEMENT OF THE PARTITION FUNCTION

In the early investigations, the solid-like degrees of freedom were repre-sented by an Einstein oscillator. Henderson [9] successfully used the Lennard-Jones and Devonshire cell model to replace the Einstein oscillator. He also formulated a partition function for rigid spheres and applied it with success. If a Lennard-Jones and Devonshire partition function is used as the solid-like partition function, then

$$f_s^N = \left[\left(\frac{2\pi mkT}{h^2} \right)^{3/2} v_f \right]^N \cdot \exp\left\{ \frac{-N\psi(o)}{2kT} \right\} \cdot (1 + n_h e^{-\varepsilon_h/kT})^N \tag{3.13}$$

where v_f, the free volume of a molecule, is given by

$$v_f = \int_{\text{cell}} \exp\left\{ \frac{-(\psi(r) - \psi(o))}{kT} \right\} dr \tag{3.14}$$

If the Lennard-Jones (6-12) potential is assumed, then

$$\psi(o) = Z\varepsilon[(1/\omega^{*4}) - (2.4/\omega^{*2})] \tag{3.15}$$

where $\omega^* = V_s/N\sigma^3$ and the effect of non-nearest neighbors as well as of the Z nearest neighbors has been taken into account.

Substituting Eqs. (3.8) and (3.13) into (3.2) and using Stirling's approximation gives

$$f_N = \left(\frac{2\pi m k T}{h^2}\right)^{(3/2)N}\left[\exp\left(-\frac{\psi(\mathrm{o})}{2kT}\right)\left(1 + n\,\frac{V - V_s}{V_s}\right)\exp\left(-\frac{\varepsilon_h}{RT}\right)v_f\right]^{N\frac{V_s}{V}}$$
$$\times \left(\frac{eV}{N}\right)^{N(V - V_s)/V} \qquad (3.16)$$

For the system of rigid spheres

$$\psi(r) = 0 \qquad r > \sigma$$
$$= \infty \qquad r < \sigma$$

and, therefore, $\psi(\mathrm{o}) = 0$ and $\varepsilon_h = 0$.

By means of machine calculations [10, 11], a transition from a solid phase to a fluid phase at about $V = 1.5V_0$ is found where $V_0 = N\sigma^3/\sqrt{2}$ is the close packed volume of the system. Accordingly, $V_s = 1.5V_0$ was used in this calculation. The values of v_f for $V_s = 1.5V_0$ were obtained by interpolating the results of Buehler *et al.* [12] and also $n = 12$ was taken. Then, the partition function for a system of rigid spheres becomes

$$f = \left\{\left(\frac{2\pi m k T}{h^2}\right)^{3/2}\frac{eV}{N}\right\}^{N}\left\{0.00967\left(8 - 11\left(\frac{V_0}{V}\right)\right)\right\}^{(3/2)N(V_0/V)} \qquad (3.17)$$

Some of the calculated results are shown in Figures 3-3 and 3-4. In this calculation, the following expressions were used:

$$\frac{PV}{NkT} = 1 - \frac{3}{2}\left(\frac{V_0}{V}\right)\ln\left\{0.00967\left[8 - 11\left(\frac{V_0}{V}\right)\right]\right\} + \frac{3}{2}\frac{11(V_0/V)^2}{8 - 11(V_0/V)} \qquad (3.18)$$

from the relation $p = -(\partial A/\partial V)_T$

$$\frac{S^E}{Nk} = \frac{3}{2}\frac{V_0}{V}\ln\left\{0.00967\left(8 - 11\,\frac{V_0}{V}\right)\right\}$$
$$- \ln\left\{1 - \frac{3V_0}{V}\ln\left[0.00967\left(8 - 11\,\frac{V_0}{V}\right)\right] + \frac{3}{2}\frac{11(V_0/V)^2}{8 - 11(V_0/V)}\right\} \qquad (3.19)$$

The theory provides an illuminating picture of the nature of the system of hard spheres that is in excellent agreement with the corresponding results of the machine calculations.

Other efforts have been made to improve the partition function. In one, the dense gas or liquid under high pressure is treated. The Einstein oscillator is modified to provide for the gradual transition from a parabolic to a square well potential [13], with success. As shown in the next chapter, the calculated critical pressure is in all cases too high. This is due, at least in part, to the neglect of clustering contributions of vacancies in the partition function. Calculations which include the dimer term [14] improve the result. These results will be discussed in detail in the next chapter.

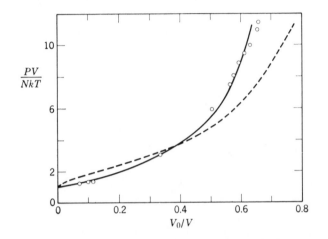

Figure 3-3 Compressibility factor for a system of rigid spheres. The solid line gives the results of the present calculations while the dotted line gives the results of the cell model and the points give the results of the machine calculations of Alder and Wainwright. (after Henderson [9])

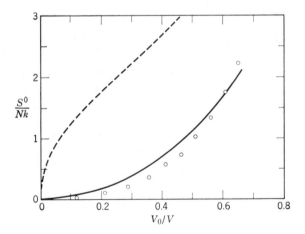

Figure 3-4 Excess entropy for a system of rigid spheres. The solid line gives the results of the present calculations while the dotted line gives the results of the cell model and the points give the results of the machine calculations of Alder and Wainwright. (after Henderson [9])

3.6 COMPARISON WITH OTHER MODELS

We next consider the reduced equations of state for various models and compare the thermodynamic properties calculated for simple liquids. The methods of obtaining the partition functions and the numerical results for other models are omitted here since we described them in the previous chapter. According to significant structure theory, the partition function [15] for simple liquids is:

$$f_N = \left[e^{E_s/RT} \left(\frac{T}{\theta} \right)^3 (1 + n_h e^{-\varepsilon/RT}) J(T) \right]^{N\frac{V_s}{V}}$$

$$\times \left[\left(\frac{2\pi mkT}{h^2} \right)^{3/2} (V - V_s) J(T) \right]^{N\frac{V-V_s}{V}} \left[\frac{N(V - V_s)}{V} \,! \right]^{-1} \quad (3.20)$$

The first set of brackets represents the solid-like partition function, and the classical approximation to the Einstein partition function was used:

$$\frac{1}{(1 - e^{-\theta/T})^3} \simeq \left(\frac{T}{\theta} \right)^3 \quad \text{where} \quad \theta < T \quad (3.21)$$

The remaining portion of Eq. (3.20) is for the gas-like degrees of freedom. Here, $J(T)$ is the partition function for the internal degrees of freedom; E_s is the configurational energy and can be written [16, 17] by applying the Lennard-Jones (6-12) potential:

$$E_s = \frac{Z}{2} N\varepsilon \left[2.4090 \left(\frac{V_s}{N\sigma^3} \right)^{-2} - 1.0109 \left(\frac{V_s}{N\sigma^3} \right)^{-4} \right] \quad (3.22)$$

At the nearest neighbor distance a' for the solid, the interaction energy $\psi(r)$ given by Eq. (2.14) attains its minimum value. Thus, we have the relation

$$a' = 2^{1/6}\sigma \quad (3.23)$$

By using Eq. (3.23) for hexagonal close packing, we obtain

$$V_s = \frac{Na'^3}{\sqrt{2}} = N\sigma^3 \quad (3.24)$$

The Einstein characteristic temperature θ in the partition function may be written as

$$\theta = \frac{h\nu}{k} = \frac{h}{k} \frac{1}{2\pi} \left(\frac{b}{m} \right)^{1/2} \quad (3.25)$$

where a simple harmonic oscillator has been assumed and the force constant,

b, can be approximated using the L.J. (6-12) potential as

$$b \frac{r^2}{2} \simeq Z(\psi(r) - \psi(a'))$$ (3.26)

Hence [18]

$$b = 2Z\varepsilon \left[22.106 \left(\frac{N\sigma^3}{V_s} \right)^4 - 10.559 \left(\frac{N\sigma^3}{V_s} \right)^2 \right] \times \frac{1}{2^{1/3}\sigma^2} \left(\frac{N\sigma^3}{V_s} \right)^{2/3}$$ (3.27)

The values of n and a for simple liquid are taken from Eqs. (3.10) and (3.12), respectively.

By substituting Eqs. (3.10), (3.12), (3.22), (3.24), and (3.25) into Eq. (3.20) and applying Stirling's approximation to $[N(V - V_s)/V]!$ we obtain

$$f = \left[\frac{(2\pi mkT)^{3/2}}{h^3} \frac{eV}{N} J(T) \right]^N \frac{e^{8.388/T^*}}{eV^*} \left(\frac{T^*}{35.01} \right)^{3/2}$$

$$\times \left[1 + 10.7(V^* - 1) \exp \left(-\frac{0.0436}{T^*(V^* - 1)} \right) \right]^{N/V^*}$$ (3.28)

where T^* and V^* are the reduced temperature and volume, respectively, and are given by the equations

$$T^* = \frac{kT}{\varepsilon}, \qquad V^* = \frac{V}{N\sigma^3}$$ (3.29)

Using Eq. (3.28), we obtain the pressure, excess* reduced entropy (S^E/Nk), and excess reduced energy ($E^E/N\varepsilon$) expressions:

$$\frac{PV}{NkT} = 1 - \frac{8.388}{V^*T^*} + \frac{\ln V^*}{V^*} - \frac{3}{2V^*} \ln \frac{T^*}{35.01} - \frac{1}{V^*}$$

$$\times \ln \left[1 + 10.7(V^* - 1) \exp - \left(\frac{0.0436}{T^*(V^* - 1)} \right) \right]$$

$$+ \frac{10.7}{T^*(V^* - 1)} \frac{0.436 + T^*(V^* - 1)}{10.7(V^* - 1) + \exp(0.436/T^*(V^* - 1))}$$ (3.30)

$$\frac{S^E}{Nk} = \frac{1}{V^*} \ln \left[1 + 10.7(V^* - 1) \exp - \left(\frac{0.0436}{T^*(V^* - 1)} \right) \right]$$

$$+ \frac{(0.4665/T^*V^*) \exp - (0.0436/T^*(V^* - 1))}{1 + 10.7(V^* - 1) \exp - (0.0436/T^*(V^* - 1))}$$

$$+ \frac{1.5}{V^*} \ln \left(\frac{T^*}{35.01} \right) + \frac{0.5}{V^*} - \frac{\ln V^*}{V^*}$$ (3.31)

---

* The excess of a property is defined as the difference between the real liquid property and the ideal liquid property. See Eq. (4.32) for definitions of excess properties.

$$\frac{E^E}{N\varepsilon} = -\frac{8.388}{V^*} + \frac{15T^*}{V^*} + \frac{(0.4665/V^*)\exp - [0.0436/T^*(V^* - 1)]}{1 + 10.7(V^* - 1)\exp - (0.0436/T^*(V^* - 1))}$$

$$(3.32)$$

In Figures 3-5 and 3-6, we compare the reduced experimental vapor pressure ($P^* = P\sigma^3/\varepsilon$) and volume ($V^*$) for simple liquids such as Ne, Ar, N_2, and CH_4 with the calculated values from the three theories. The agreement

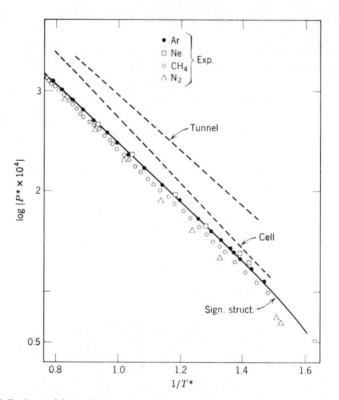

Figure 3-5 Logarithm of the reduced vapor pressure, P^*, vs. $1/T^*$ for simple liquids. The experimental values have been reduced by using ε/k and $N\sigma^3$. The significant structure theory (full curve) is compared with the cell theory and the tunnel theory (broken curves). (after Ree *et al.* [15])

between significant structure theory and experiment is very good and is better than the values from the other theories.

According to the significant structure approach, the reduced vapor pressure and volumes are obtained by plotting the Helmholtz free energy versus reduced volume as shown in Figure 3-7. Now, the melting temperature is defined as the temperature at which the slope of the common tangent between solid and liquid minimum (in Figure 3-7) is $P^* = \sigma^3/\varepsilon$, i.e., $P = 1$ atmosphere. In Table 3-2, the melting properties for the significant structure theory are

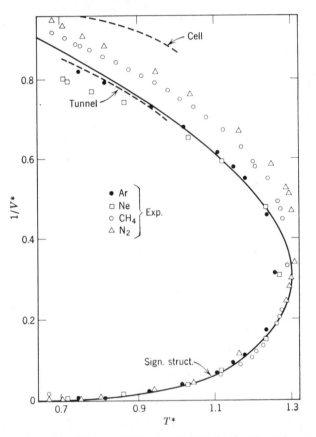

Figure 3-6 Reduced density vs. T^*. The experimental values have been reduced by using ε/k and $N\sigma^3$. The significant structure theory (full curve) is compared with the cell theory and the tunnel theory (broken curves). (after Ree *et al.* [15])

compared with experimental data for liquid argon, and also with those of the other theories at $T^* = 0.70$. In columns 4 and 5 of Table 3-2, the calculated reduced excess entropy and excess energy at the melting point are also compared. The cell theory is a theory of superheated solids rather than a theory of liquids, as can be seen by comparing the data for solid argon with values calculated from the cell theory.

Table 3-2   Theoretical and Experimental Properties at the Melting Point

	Melting Temp. T^*	Reduced Volume V^*	Reduced Excess Entropy S^E/Nk	Reduced Excess Energy $E^E/N\varepsilon$
Significant Structure Theory	0.711	1.159	−3.89	−6.19
Tunnel Theory		1.184	−4.8	−5.9
Liquid Argon	0.701	1.178	−3.64	−5.96
L.J. and D. Theory		1.037	−5.51	−7.32
Solid Argon	0.701	1.035	−5.53	−7.14

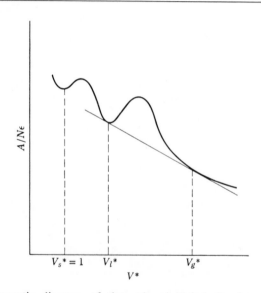

Figure 3-7   Schematic diagram of the reduced Helmholtz free energy, $A/N\varepsilon$, plotted against reduced volume V^* at a constant reduced temperature T^*. Here, V_s^*, V_l^*, and V_g^* are reduced volumes of the solid, liquid, and gas respectively. (after Ree *et al.* [15])

The critical constants are obtained by using the conditions

$$\left(\frac{\partial P}{\partial V}\right)_T = 0 \quad \text{and} \quad \left(\frac{\partial^2 P}{\partial V^2}\right)_T = 0$$

In Table 3-3 we compare the critical properties for the three theories and experimental data for simple liquids. It can be seen that the significant

Table 3-3 Experimental and Theoretical Critical Constants

	T_c^*	P_c^*	V_c^*	(PV/RT)
Mean Values for Ne, Ar, N_2, CH_4	1.277	0.121	3.09	0.292
Significant Structure Theory	1.306	0.141	3.36	0.362
Tunnel Theory	1.07	0.37	1.8	0.6
L.J. and D. Theory	1.30	0.434	1.77	0.591

structure theory is better than the other theories. The tunnel theory gives slightly worse agreement with experiment than the cell theory. Almost certainly this is due to the approximation of treating the motion in the tunnels as strictly one-dimensional.

3.7 SOME THEORETICAL DISCUSSIONS OF SIGNIFICANT STRUCTURE THEORY

Significant structure theory has been successfully applied to many liquids, as will be shown in later chapters. The theory is based on the idea that the vapor is mirrored in the liquid as vacancies which transform solid-like into gas-like degrees of freedom. When properly formulated, such a model should be a useful description of what actually happens, as in fact it is.

The usefulness of the model cannot be doubted. However, criticisms have been raised against the theory. One of its disadvantages is that "it has not been derived from an exact partition function by any mathematically well-defined approximation, but is a result of intuition." The cell theory has been given a statistical mechanical foundation by Kirkwood [19]. It is, therefore, instructive to trace the correlation between these two models. Several attempts [20–22] have been made. One of these is pointed out by Pierotti [20]. According to the free volume theory [Section 2.4], the partition function for N identical monatomic molecules randomly distributed over L lattice cells is:

$$f_{fv} = \left(\frac{L!}{N!(L-N)!}\right) \frac{(2\pi mkT)^{(3/2)N}}{h^{3N}} (v_f)^N \cdot \exp\left(-yNZ\phi(o)/2kT\right) \quad (2.34)$$

We use the following linear assumption [23] for the relation between v_f and y:

$$v_f = yv_1 + (1-y)v_0 = yv_1\{1 + [(1-y)/y](v_0/v_1)\} \quad (3.33)$$

where v_1 and v_0 are the free volumes of molecules with y equal to unity and zero, respectively. Substitution of Eq. (3.33) into Eq. (2.34) and rearranging gives

$$f_{fv} = \left[\frac{(2\pi mkT)^{3/2}}{h^3} v_f \exp\left(-\frac{Z\phi(o)}{2kT}\right)\{(1 + (1 - y)/y)(v_0/v_1)\}\right]^{yN}$$

$$\times \left[\frac{(2\pi mkT)^{3/2}}{h^3} v_f\right]^{(1-y)N} \left(\frac{N}{L}\right)^{yN} \frac{L!}{N!(L - N)!} \quad (3.34)$$

Since the total volume of the lattice (V) is equal to Lv_0, y is equal to V_s/V where V_s is the volume of a solid-like lattice of N cells. Thus, Eq. (3.34) can be rewritten as

$$f_{fv} = \left[\frac{(2\pi mkT)^{3/2}}{h^3} v_f\left\{[1 + (V - V_s)/V_s]\left(\frac{v_0}{v_1}\right)\right\}\right]^{NV/V_s}$$

$$\times \left[\frac{(2\pi mkT)^{3/2}}{h^3}(V - V_s + V_{ex})\right]^{\frac{N(V-Vs)}{V}} (L)^{-N}(N)^{N(N/L)} \frac{L!}{N!(L - N)!}$$

$$(3.35)$$

where $V_{ex} = Nv_1$ is the effective excess volume of the solid due to the nearest neighbor interactions. If we set $V_{ex} = 0$, and identify n with v_0/v_1, the significant structure partition function differs only in the combinatorial factor. This difference arises in the formation of the partition function. The free volume theory does not have the limiting values of an ideal gas as the significant structure theory does. Hildebrand and Archer [24] criticize the quasi-crystalline models of liquids as implied in such terms as "lattice," "cells," "holes," "vacancies," and "dislocations," and claim that any sort of solid-like structure is inconsistent with experimental facts. As an exemplification of the problem, they cited p- and m-xylenes, which are very similar in the liquid state, but differ widely in their melting temperatures. m- and p-xylene boil within 0.8°C of each other, and their liquid heat capacities are almost identical within experimental error. Further, their molar volumes differ by only 0.5 cc between 100 and 13.2°C, the melting point of p-xylene. Meta-xylene exists as liquid down to −47.9°C. (Figure 3-8.) The percentages of shrinkage accompanying the solidification at the melting points are 16.8 and 8.5 for p- and m-xylenes, respectively. Noticing these facts, Hildebrand and Archer insisted that "near identity of the molar volumes and of the heat capacities of these two isomers above 13.2°C, contrasting with the enormous difference in the freezing temperature, is surely inconsistent with any content in either isomer of solid-like aggregates."

However, the so-called "inconsistent" facts have a natural explanation [25] if we use significant structure theory. The enormous difference in the melting points and the considerable difference in the fusion entropies (14.28 e.u. for p-xylene, 12.26 e.u. for m-xylene) are consistent with the fact that the

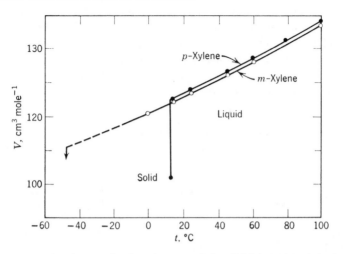

Figure 3-8 Molal volumes of xylene isomers. (after Hildebrand and Archer [24])

observed sublimation energy E_s for p-xylene (13,797 cal) is much larger than that (13,243 cal) of m-xylene. The differences between the E_s values arises from a closer, more orderly packing of solid p-xylene than of m-xylene. As a consequence, p-xylene has a larger entropy of fusion and a higher melting temperature than m-xylene. In order to explain the near identity of the thermodynamic properties above 13.2°C for the two xylenes, we assume (as was to be expected) that a change in packing occurs in the melting processes. By this transition, the molar volume of the solid-like molecules becomes larger than the observed solid molar volume for both xylenes, and the molar sublimation energies of the solid-like molecules become nearly equal. Our assumption of phase transitions at and near melting is supported by the observation [26] that an anomaly in light scattering occurs near the melting temperature of p-xylene. These ideas were introduced into the model. The following partition function was used for both xylenes.

$$f = \left\{ \frac{e^{E_s/RT}}{(1 - e^{-\theta/T})^6} \left[1 + n\left(\frac{V - V_s}{V_s}\right) \exp -\left(\frac{aE_s V_s}{(V - V_s)RT}\right) \right] \right.$$

$$\times \prod_{i=1}^{2} \frac{(8\pi^3 I_i kT)^{1/2}}{\sigma_m h} \prod_{i=1}^{46} \frac{1}{1 - e^{-h\nu_i/kT}} \right\}^{N\frac{V_s}{V}}$$

$$\times \left\{ \frac{(2\pi mkT)^{3/2}}{h^3} \frac{eV}{N} \frac{8\pi^2 (8\pi^3 ABC)^{1/2}(kT)^{3/2}}{\sigma h^3} \right.$$

$$\times \prod_{i=1}^{2} \frac{(8\pi^3 I_i kT)^{1/2}}{\sigma_m h} \prod_{i=1}^{46} \frac{1}{1 - e^{-h\nu_i/kT}} \right\}^{N\frac{V - V_s}{V}} \qquad (3.36)$$

For the formulation of the partition function, free rotation of the methyl group in the benzene ring and no rotation of the solid-like molecules was assumed. Here, I_i is the reduced moment of inertia of the CH_3 group, and the other quantities have been defined previously. Equation (3.36) yields a satisfactory explanation of all thermodynamic and transport properties of both xylenes. Some of the results are shown in Table 3-4.

Table 3-4   Molar Volumes of Xylenes

	p-xylene			m-xylene		
T (°K)	V_{calc} (cc)	V_{obs} (cc)	$\Delta\%$	V_{calc} (cc)	V_{obs} (cc)	$\Delta\%$
225.29 (m.p)	—	—	—	115.23	115.23	0.00
286.39 (m.p)	122.90	122.5	0.33	—	—	—
298.16	124.04	—	—	120.31	123.47	−2.56
313.16	125.44	125.8	−0.29	121.59	125.35	−2.99
353.16	129.60	131.3	−1.29	125.46	130.72	−4.02
393.16	134.68	137.6	−2.12	130.15	136.73	−4.81
411.51 (b.p)	137.40	—	—	—	—	—
412.26 (b.p)	—	—	—	132.75	139.80	−5.04

3.8   A SUMMARY OF OUR LIQUID MODEL CONCEPTS

The fluidized vacancies present in liquids should not be confused with the Frenkel defect in solids. The Frenkel locked-in vacancies no doubt occur in the liquid as well as in the solid, as do the other solid defects. However, the continually collapsing fluidized vacancies contribute substantially to the entropy of the system by (a) converting vibrational into translational degrees of freedom and (b) by conferring configurational degeneracy on those neighboring molecules which have sufficient kinetic energy to keep the vacancy open and to deny its use to competitors. This extra entropy makes fluidized vacancies overwhelmingly more plentiful than the usual defects of solids.

In his book *The Kinetic Theory of Liquids* [27] Frenkel sets forth his ideas of the liquid state. His considerations are qualitative, as he never seriously undertook a quantitative treatment. The degree of agreement with our point of view as expressed in our early papers is interesting. In developing the quantitative theory presented here we have been obliged to develop a wide variety of new concepts not required in the earlier qualitative considerations.

We summarize some of these concepts at this point:

(a) The liquid phase differs from a solid phase in that the kinetic energy of molecules in the liquid has become large enough that it can balance the potential energy tending to make molecules collapse into the holes that are present. The entropy gained from the appearance of holes is negligible at low concentrations of the holes but increases cooperatively with these holes becoming fluidized so that the Helmholtz free energy at any temperature goes through a minimum with volume near the melting point.

(b) The distribution of holes in a liquid may be considered to have the average volume of a vacancy, are mobile, and are called fluidized vacancies.

(c) The fluidized vacancies in a liquid mirror the molecules in the vapor in concentration and in behavior.

(d) In a mole of liquid there are $(V - V_s)/V_s$ moles of fluidized vacancies each of which confers gas-like properties on V_s/V molecules. The quantity V_s/V is the fraction of positions next to a vacancy which are occupied by molecules. Thus, we have the fraction

$$\frac{V_s}{V} \frac{(V - V_s)}{V_s} = \frac{V - V_s}{V}$$

of our degrees of freedom acting like gas-molecules and the remainder V_s/V are solid-like. V and V_s are the molal volumes of liquid and solid respectively.

(e) The mean value of a liquid property, X, is accordingly:

$$X = X_s \frac{V_s}{V} + X_g \frac{V - V_s}{V}$$

where X_s and X_g are the value of the property in the solid and vapor states, respectively.

(f) The vacancies, in addition to introducing gas-like properties, give the solid-like molecules a degeneracy $(1 + n_h e^{-\varepsilon_h/RT})$. Here n_h is the number of vacancies made available to a molecule through the expenditure of an energy, ε_h/N.

(g) By considering the general form that n_h and ε_h must have according to our model, and using the fact that at the melting temperature the liquid approximates a lattice, we find that for simple spherical molecules, like argon, $n_h = 10.7(V - V_s)/V_s$ and $\varepsilon_h = 0.0054E_s V_s/(V - V_s)$. For other types of molecules the numerical constants in n_h and ε_h differ in a predictable manner. The success of the theory leaves little to be desired.

REFERENCES

[1]   H. Eyring, T. Ree, and N. Hirai, *Proc. Natl. Acad. Sci. (U.S.)*, **44**, 683 (1958).
      H. Eyring and T. Ree, *ibid.*, **47**, 526 (1961).
[2]   H. Eyring and R. P. Marchi, *J. Chem. Edu.*, **40**, 562 (1963).

[3] H. Eyring and T. Ree, *Bulletin of the New Mexico Academy of Science* (1964).
[4] T. S. Ree, T. Ree, and H. Eyring, *Angew. Chem. Internat. Edit.*, Vol. 4, No. 11, 923 (1956).
[5] M. E. Zandler and M. S. Jhon, *Ann. Rev. Phys. Chem.*, **17**, 373 (1966).
[6] A. Eisenstein and N. Gingrich, *Phys. Rev.*, **62**, 261 (1942).
[7] J. Walter and H. Eyring, *J. Chem. Phys.*, **9**, 393 (1941).
[8] E. J. Fuller, T. Ree, and H. Eyring, *Proc. Natl. Acad. Sci. (U.S.)*, **45**, 1594 (1959).
[9] D. Henderson, *J. Chem. Phys.*, **39**, 1857 (1963).
[10] W. W. Wood and J. D. Jacobson, *J. Chem. Phys.*, **27**, 1207 (1957).
[11] B. J. Alder and T. E. Wainwright, *J. Chem. Phys.*, **27**, 1208 (1957).
[12] R. J. Buehler, R. H. Wentorf, Jr., J. O. Hirschfelder, and C. F. Curtiss, *J. Chem. Phys.*, **19**, 61 (1951).
[13] T. S. Ree, T. Ree, and H. Eyring, *Proc. Natl. Acad. Sci. (U.S.)*, **48**, 501 (1962).
[14] J. Grosh, M. S. Jhon, T. Ree, and H. Eyring, *Proc. Natl. Acad. Sci. (U.S.)*, **57**, 1566 (1967).
[15] T. S. Ree, T. Ree, H. Eyring, and R. Perkins, *J. Phys. Chem.*, **69**, 3322 (1965).
[16] J. E. Lennard-Jones and A. F. Devonshire, *Proc. Roy. Soc. (London)*, **A163**, 53 (1937).
[17] R. H. Wentorf, R. J. Buehler, J. O. Hirschfelder, and C. F. Curtiss, *J. Chem. Phys.*, **18**, 1484 (1950).
[18] I. Prigogine, *The Molecular Theory of Solutions*, North Holland Publishing Co., Amsterdam, 1957, p. 130.
[19] J. G. Kirkwood, *J. Chem. Phys.*, **18**, 380 (1950).
[20] R. A. Pierotti, *J. Chem. Phys.*, **43**, 1072 (1965).
[21] T. S. Ree, T. Ree, and H. Eyring, *Proc. Natl. Acad. Sci. (U.S.)*, **51**, 344 (1964).
[22] M. S. Jhon, *J. Korean Chem. Soc.*, **11**, 60 (1967).
[23] D. Henderson, *J. Chem. Phys.*, **37**, 631 (1962).
[24] J. H. Hildebrand and G. Archer, *Proc. Natl. Acad. Sci. (U.S.)*, **47**, 1881 (1961).
[25] M. S. Jhon, J. Grosh, T. Ree, and H. Eyring, *Proc. Natl. Acad. Sci. (U.S.)*, **54**, 1419 (1965).
[26] G. S. Kastha, *Indian J. Phys.*, **32**, 473 (1958).
[27] J. Frenkel, *Kinetic Theory of Liquids*, Dover, New York, 1955.

chapter 4

THE THERMODYNAMICS
OF LIQUIDS

4.1 INTRODUCTION

In previous chapters we developed the significant structure theory of liquids and made comparisons and correlations with other liquid theories exemplifying simple liquids and hard spheres. In this chapter we describe the evaluation of thermodynamic properties of various liquids from significant structure theory.

Using the basic partition function given in Eq. (3.2), we are able to evaluate thermodynamic properties from the melting point to the critical point for a wide variety of substances, from simple liquids such as argon to complicated liquid systems such as water and various mixtures. To do this, values of E_s, V_s, θ, a, n, and other physical constants must be either known or calculated. Methods of obtaining such quantities were discussed in the previous chapter. The Helmholtz free energy, A, is

$$A = -kT \ln f \qquad (4.1)$$

Since we now have A as a function of volume and temperature, we can calculate all thermodynamic properties by the following procedure. For simplicity, we first consider systems with a single component; binary liquid systems will be described later.

(a) Vapor Pressures, Molar Volume, and Critical Properties

If the Helmholtz free energy A is plotted as a function of volume at a constant temperature, as shown in Figure 4-1, and if a common tangent to the points corresponding to the liquid and vapor phases is drawn, the vapor pressure is given by the slope of the common tangent and the abscissas of the two points indicate the respective volume of the liquid and vapor. The critical temperature, T_c, the critical pressure, P_c, and the critical volume, V_c, are obtained by applying the usual conditions:

$$\left(\frac{\partial P}{\partial V}\right)_T = 0; \quad \left(\frac{\partial^2 P}{\partial V^2}\right)_T = 0; \quad P = -\left(\frac{\partial A}{\partial V}\right)_T \tag{4.2}$$

which are evaluated by using Eq. (3.2).

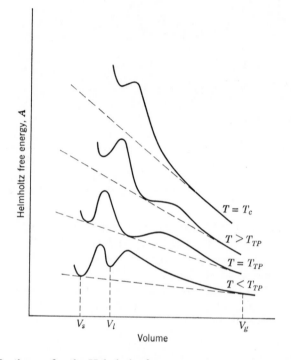

Figure 4-1 Isotherms for the Helmholtz free energy, A, vs. volume. The scales are greatly distorted. T_c is the critical temperature and T_{TP} is the triple point temperature. (after Thomson *et al.* [5])

(b) Other Thermodynamic Properties

These are given by the following equations, which are evaluated using Eq. (3.2):

$$S = -\left(\frac{\partial A}{\partial T}\right)_V = k\left(\frac{\partial (T \ln f)}{\partial T}\right)_V$$

$$E = -T^2\left(\frac{\partial (A/T)}{\partial T}\right)_V = kT^2\left(\frac{\partial \ln f}{\partial T}\right)_V$$

$$H = E + PV = kT^2 \left(\frac{\partial \ln f}{\partial T}\right)_V + VkT\left(\frac{\partial \ln f}{\partial V}\right)_T$$

$$G = A + PV = -kT \ln f + VkT\left(\frac{\partial \ln f}{\partial V}\right)_T$$

$$C_V = \left(\frac{\partial E}{\partial T}\right)_V = k\left(\frac{\partial}{\partial T}\left(\frac{T^2 \partial \ln f}{\partial T}\right)\right)_V$$

$$\alpha = \frac{1}{V}\left(\frac{\partial V}{\partial T}\right)_P = -\frac{(\partial P/\partial T)_V}{V(\partial P/\partial V)_T} = -\frac{1}{V}\frac{[\partial/\partial T(kT)(\partial \ln f/\partial V)_T]_V}{kT(\partial^2 \ln f/\partial V^2)_T}$$

$$\beta = -\frac{1}{V}\left(\frac{\partial V}{\partial P}\right)_T = -\frac{1}{V}\frac{1}{kT(\partial^2 \ln f/\partial V^2)_T}$$

$$C_P = C_V + \alpha^2 TV/\beta$$

(4.3)

4.2 INERT GASES

For the liquid inert gases, we use the partition function given in Eq. (3.9): E_s, θ, and V_s are taken from the properties of the solid phase; n and a are determined from Eqs. (3.10b) and (3.12b).

The calculated results were summarized in Chapter 3 (Section 3.4 and Table 3-1). The agreement between experiment and theory is excellent except that the calculated critical pressure is in all cases high. This discrepancy arises from the neglect of the clustering of vacancies, paralleling the neglect of molecular clustering in the vapor. The concentrations of clusters become significant as the critical point is approached and thus should be taken into account in the partition function. Recently Grosh, Jhon, Ree, and Eyring [1] introduced this idea into the partition function, and improved the result. The partition function for the gas-like degree of freedom f_g' that they use is

$$f_g' = f_g e^{NK/V} \tag{4.4}$$

Table 4-1 Calculated and Observed Critical Constants [1]

	T_c (°K)			P_c (atm.)			V_c (cc)		
	Calc. I	Calc. II	Obs.	Calc. I	Calc. II	Obs.	Calc. I	Calc. II	Obs.
Ar	149.5	150.0	150.7	52.9	50.8	48.0	83.5	86.0	75.3
Kr	209.5	210.0	210.6	62.4	60.5	54.24	101.0	103.0	—
Xe	292.4	293.2	289.8	69.6	67.8	58.2	125.0	127.0	113.8

Calc. I denotes the calculation using perfect gas approximations while Calc. II shows the calculation using the dimer correction where f_g represents the ideal gas partition function and K is the quotient of the partition functions for dimers and monomers. Some of the results are summarized in Table 4-1. Calculations of the critical pressure P_c are considerably improved by even this modest attempt to take account of clustering.

4.3 INORGANIC LIQUIDS

Significant structure theory has been applied to a number of inorganic liquids: nitrogen [2], oxygen [3], chlorine [4], fluorine [5], bromine [6], iodine [6], hydrides [7], hydrazine [8], diborane [8], hydrogen halides [9], carbon dioxide [10], carbon disulfide [10], and carbon oxysulfide [10]. The theory is being presently applied to complicated boron hydrides [11] such as pentaborane with success.

Let us examine the application of the theory to some typical liquids. First, we consider liquid oxygen [3]. Oxygen is a diatomic molecule with a normal entropy of fusion (1.95 e.u.) and also has two first-order transitions in the solid state (one at 23.66°K, the other at 43.76°K). This suggests that an oxygen molecule rotates in the solid state near the melting point. Actually, in 1952, Crawford *et al.* found rotational bands in Raman spectra in liquid oxygen. Therefore, the rotational term with a triplet ground state is included in both the gas-like and solid-like parts of the partition function. For the construction of the partition function, the dimer term has been ignored since the dimer concentration is very small. The partition function for liquid oxygen is then

$$f_{O_2} = \left\{ \frac{e^{E_s/RT}}{(1 - e^{-\theta/T})^3} \left(1 + n(x - 1) \exp \left(- \frac{aE_s}{(x - 1)RT} \right) \right) \right.$$
$$\left. \cdot \frac{3(8\pi^2 I k T)}{2h^2} \frac{1}{1 - \exp(-hv/kT)} \right\}^{N\frac{V_s}{V}}$$
$$\cdot \left\{ \frac{(2\pi m k T)^{3/2}}{h^3} \frac{eV}{N} \frac{3(8\pi^2 I k T)}{2h^2} \frac{1}{1 - \exp(-hv/kT)} \right\}^{N\frac{V - V_s}{V}} \tag{4.5}$$

Here, $x = V/V_s$, E_s, θ, V_s, m, N, n, and a are as defined previously; v and I are the internal molecular vibrational frequency and the moment of inertia, respectively. Comparison of some of the calculated and observed properties are shown in Table 4-2. The parametric values used are: $n = 11.89$, $\theta = 56.02$, $V_s = 24.14$ cc/mole, $a = 0.5893 \times 10^{-4}$, $E_s = 1808$ cal/mole. The results are satisfactory.

Next, we examine liquid diborane. Its abnormally high entropy of fusion (9.859 e.u.) and the absence of solid state transitions indicate that the diborane molecule does not rotate freely in the solid state. Also, the increase in volume upon melting is not sufficient to permit the rotation of solid-like molecules in the liquid. Accordingly, the rotational term appears only in the

Table 4-2 Properties at the Triple, Boiling, and Critical Points for Liquid Oxygen

T (°K)	P (atm)	V (cc/mole)	S (e.u./mole)	
54.36 (T_t)	0.00146	24.37	16.62	Calc.
54.36	0.00150	24.37	16.04	Obs.
0.00	−2.67	0.00	3.63	Δ (%)
(90.21)	1.081	27.58	23.07	Calc.
90.21 (T_b)	1.000	28.06	22.50	Obs.
	8.10	−1.67	2.53	Δ (%)
162.78	60.55	79.71	28.84	Calc.
154.34 (T_c)	49.713	74.44	—	Obs.
5.47	21.81	7.08	—	Δ (%)

gas partition function and a six-degree Einstein oscillator term appears in the solid. The partition function for diborane [4] accordingly takes the following form:

$$f = \left\{ \frac{e^{E_s/RT}}{(1 - e^{-\theta/T})^6} \prod_{i=1}^{18} \frac{1}{1 - e^{-hv_i/kT}} \left[1 + n(x - 1)e^{-aE_s/(x-1)RT}\right] \right\}^{N\frac{V_s}{V}}$$

$$\times \left\{ \frac{(2\pi mkT)^{3/2}}{h^3} \frac{eV}{N} \frac{8\pi^2(8\pi^3 ABC)^{1/2}(kT)^{3/2}}{4h^3} \prod_{i=1}^{18} \frac{1}{1 - e^{-hv_i/kT}} \right\}^{N\frac{V-V_s}{V}} \quad (4.6)$$

Here A, B, and C are the moments of inertia, and the other quantities have been defined previously. The parametric values we used are: $V_s = 56.81$ cc/mole; $\theta = 51.77$; $a = 0.136 \times 10^{-4}$; $E_s = 4344.9$ cal/mole and $n = 11.94$. Some of the results are shown in Tables 4-3, 4-4, and 4-5 and in Figure 4-2, and the excellent agreement between theory and experiment is evident.

Table 4-3 Calculated and Observed Molar Volume and Vapor Pressure of Liquid Diborane

Temperature (°K)	V_{calc} (cc/mole)	V_{obs} (cc/mole)	Δ%	P_{calc} (atm.)	P_{obs} (atm.)	Δ%
108.3 (m.p)	57.09	(57.09)	0.00	0.708×10^{-3}	0.710×10^{-3}	−0.28
130.12	58.96	—	—	0.01706	0.01690	0.95
145.03	60.49	—	—	0.08082	0.07982	1.25
155.40	61.70	—	—	0.1947	0.1910	1.94
165.14	62.95	60.99	−3.21	0.3953	0.3910	1.10
175.04	64.36	62.50	−2.97	0.7387	0.7280	1.47
180.66 (b.p)	65.22	63.36	−2.93	1.0163	1.0000	1.63

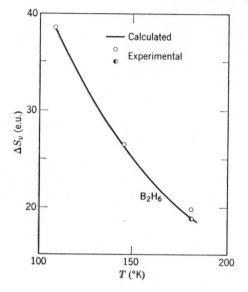

Figure 4-2 Entropies of vaporization of liquid B_2H_6 vs. temperature. (after Jhon *et al.* [8])

Table 4-4 Calculated Values for α, β, C_v, and C_p for Liquid Diborane

Temperature (°K)	$\alpha \times 10^3$ (deg^{-1})	$\beta \times 10^5$ (atm^{-1})	C_v (cal/mole deg)	C_p (cal/mole deg)	C_{pobs} (cal/mole deg)	$\Delta\%$
108.3 (m.p)	1.425	0.538	12.12	17.69	18.23	2.96
130.12	1.613	0.807	12.28	18.19	18.03	−0.89
145.03	1.823	1.094	12.38	18.76	18.09	3.70
155.40	1.984	1.349	12.27	19.17	18.16	5.56
165.14	2.150	1.642	12.56	19.56	18.26	7.11
175.04	2.336	2.008	12.66	19.99	18.37	8.82
180.66 (b.p)	2.449	2.253	12.72	20.23	—	—

Table 4-5 Critical Point Properties for Liquid Diborane [4]

	Calc.	Obs.	$\Delta\%$
T_c	317.6	289.9	9.55
P_c	50.25	40.85	23.01
V_c	187.8	173.1	8.49

4.4 ORGANIC LIQUIDS

The theory has been applied to a number of organic liquids ranging from simple hydrocarbons such as methane to complicated molecules such as the xylenes. So far, the following molecules have been treated successfully: methane [2, 12, 13], ethane [12, 15], propane [12], benzene [2, 12, 13], cyclohexane [19], methyl halides [14, 15], nitromethane [11], methanol [15], carbon tetrachloride [16], ethylene dichloride [17], and p- and m-xylenes [18].

The experimental heat capacity curves for both CH_4 and CCl_4 indicate that the molecules rotate in the solid state. Consequently, they must also rotate in the liquid state. The partition function for liquid CH_4 and CCl_4 has the following form:

$$f = \left\{ \frac{e^{E_s/RT}}{(1 - e^{-\theta/T})^3} \left[1 + n(x - 1) \exp\left(-\frac{aE_s}{(x - 1)RT} \right) \right] \right\}^{N\frac{V_s}{V}}$$

$$\cdot \left\{ \frac{(2\pi mkT)^{3/2}}{h^3} \cdot \frac{eV}{N} \right\}^{N(V - V_s)/V}$$

$$\cdot \left\{ \prod_{i=1}^{9} \frac{1}{1 - e^{-hv_i/kT}} \cdot \frac{8\pi^2(8\pi^3 ABC)^{1/2}(kT)^{3/2}}{12h^3} \right\}^N \qquad (4.7)$$

All symbols have been defined. Since the partition functions for the internal vibrations and the rotations occur in both portions of the partition function, we have factored them out for convenience. Some of the results for methane and carbon tetrachloride are given in Table 4-6.

The criteria for free rotation or nonrotation of a solid-like molecule are frequently ambiguous. Often it is more reasonable to assume hindered rotational degrees of freedom for the solid-like molecule in the liquid. This has been approximated as a vibration with a frequency of $\theta k/h$. Recently, McLaughlin and Eyring [20] introduced an approximate hindered rotational partition function f_{HR} into f_s. They used as the f_s for chlorine:

$$f_s = \frac{e^{E_s/RT}}{(1 - e^{-\theta/T})^3} (f_{HR})^2 \qquad (4.8)$$

and for benzene:

$$f_s = \frac{e^{E_s/RT}}{(1 - e^{-\theta/T})^3} (f_{HR})^3 \qquad (4.9)$$

Here

$$f_{HR} = f_{vib} + (f_{FR} - f_{vib}) \exp\left[\frac{-BV_s}{RT(V - V_{so})} \right] \qquad (4.10)$$

Table 4-6   Thermodynamic Properties of Liquid CH_4 and CCl_4

	n	a	θ	V_s (cc/mole)	E_s (cal/mole)
CH_4	11.05	0.364×10^{-2}	71.34	31.06	2201
CCl_4	11.64	0.479×10^{-3}	53.53	89.39	4334

	CH_4 [13]		CCl_4 [16]	
	Calc.	Obs.	Calc.	Obs.
T_m (°K)	90.65	90.65	250.22	250.22
V_m (cc)	33.63	33.63	92.12	92.12
P_m (atm)	0.1154	0.1154	0.0111	0.0111
ΔS_v (e.u.)	23.20	23.20	—	—
T_b (°K)	111.67	111.67	349.9	349.9
V_b (cc)	36.12	37.79	102.02	103.89
P_b (atm)	0.9954	1.0000	1.0317	1.0000
ΔS_b (e.u.)	17.514	17.51	—	—
T_c (°K)	211.29	191.05	567.3	556.3
P_c (atm)	60.59	45.8	56.85	45
V_c (cc)	103.83	99.34	295.0	275.7
ΔS_c (e.u.)	—	—	21.22	20.41

where f_{FR} is a free rotational partition function, and f_{vib} is the restricted vibrational partition function which they approximated as unity. V_{so} is the solid volume at which a free rotation starts, and the constant B divided by the number of vacancies is taken as the barrier height. Using Eq. (4.10) for the solid-like partition function, they successfully calculated the properties of N_2, Cl_2, CH_4, CH_3Cl, and CCl_4.

4.5   MOLTEN METALS

Since the conducting (non-localized) electrons of a metal are spread out over the positive ions, it is the much smaller positive metal ions, rather than the atoms, which are unlocked from their lattice positions by a correspondingly smaller volume increase upon melting. Usually, the ions are only about a third as large as the metal atoms, so that the vacancies introduced in melting are correspondingly smaller. Thus when metals melt, they expand only 3 or 4%, about a third as much as a normal liquid such as argon. However, the entropy of melting is almost the same as that of a normal liquid, so n must have a value about three times that of a normal liquid. Any hole theory for liquid metals should incorporate this factor of relatively small hole sizes required for a degeneracy. For the construction of the partition function of a liquid metal [21], an Einstein solid is assumed for the solid-like degree of freedom, and monatomic and diatomic terms are included in the gas-like part of the partition function, since an appreciable concentration of the dimer exists in the vapor. The application of the method of significant structures to metals takes the following form:

$$f = \left\{ \frac{e^{E_s/RT}}{(1 - e^{-\theta/T})^3} \left[1 + n \frac{V - V_s}{V_s} \exp \left(-\frac{aE_s V_s}{RT(V - V_s)} \right) \right] \right\}^{N \frac{V_s}{V}}$$

$$\cdot \left\{ \left(f_1 \frac{eV}{N} \right)^2 + \left(f_2 \frac{eV}{N/2} \right) \right\}^{\frac{N}{2} \frac{V - V_s}{V}}$$

where

$$f_1 = \frac{g(2\pi m_1 kT)^{3/2}}{h^3}$$

$$f_2 = \frac{(2\pi m_2 kT)^{3/2}}{h^3} \cdot \frac{8\pi^2 IkT}{2h^2} \cdot \frac{e^{D/RT} - 1}{1 - e^{-hv/kT}} \tag{4.11}$$

and m_1, m_2, I, D, v, and g are, respectively, the masses of the monatomic and diatomic metal atoms, the moment of inertia, the dissociation energy, the ground state vibrational frequency of the dimer, and the electronic degeneracy factor of the metal atoms; the other quantities have been defined previously. As the temperature approaches the critical point, the atoms no longer behave as if they were ions, since the electrons are no longer so completely smeared over the positive ions; accordingly, the values of n must approach the value for a normal liquid.

Equation (4.11) has been applied to calculate the properties of liquid Na, Hg, Cu, and Pb. Some typical results are given in Figure 4-3 in which the molar volume is plotted against the temperature for liquid sodium. The maximum error for a temperature range of 700°C is less than 3%.

4.6 FUSED SALTS

Relatively little has been done to develop the theory of fused salts, in spite of their present practical and theoretical importance. Carlson, Eyring, and Ree [22] applied the theory of significant structure successfully to fused salts. Blomgren [23] also applied an extended significant structure theory to fused salts. Other successful approaches are that of McQuarrie [24], using the methods of Lennard-Jones and Devonshire [25], and that of Kirkwood, Bogolyubov, and Green [26], who used molecular distribution functions.

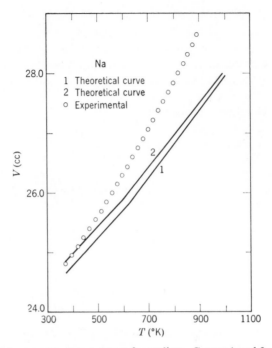

Figure 4-3   Volume versus temperature for sodium. Curves 1 and 2 were calculated using values of $n = 25$, $a = 0.002500$ and $n = 30$, $a = 0.006000$, respectively. (after Carlson *et al.* [21])

The above treatments are all restricted to a consideration of the highly ionized molten alkali halides. With other molten salts, the matter is often more complicated. Mercuric halide exemplifies the slightly ionized salts. Here, the equilibrium $HgX_2 \rightleftharpoons HgX^+ + HgX_3^-$ is established, with a strong preponderance of neutral molecules [27]. Jhon, Clemena, and Van Artsdalen [28] recently developed significant structure theory to treat this type of problem.

We will first describe the theory of the highly ionized systems. The observed percentage change in volume of argon upon melting is 12%. However, for the molten alkali halides, the expansion [29] is approximately twice that of argon, as can be seen in Table 4.7. This is an expected result since the entropy of

Table 4-7   Volume Change at the Melting Point [29]

	NaCl	NaBr	NaI	KCl	LiCl
Volume expansion	25.0	22.4	18.6	17.2	26.2

melting comes from the randomness introduced by the excess volume, and only half the extra space is accessible to positive and half to negative ions. Thus, twice the percentage expansion is required to get the same entropy increase for each kind of ion. Since only half the excess volume $(V - V_s)$ provides vacancies for each kind of ion, we expect the factor n in the positional degeneracy term to be only about half as large for salts as it is for argon, as is the case. In constructing the partition function, the solid-like portion of the partition function is raised to the power $2N(V_s/V)$, since a mole of alkali halide contains $2N$ particles. A common Einstein temperature is used for both ions as an approximation; E_s/N is the sublimation energy per alkali halide molecule, and must be divided by two to get the required E_s per ion in the partition function. To account for the long range nature of the coulombic interionic potential, the solid E_s has been multiplied by the factor $(V/V_s)^{1/3}$. The gas-like significant structure is composed of diatomic alkali halide molecules. These considerations lead to the following partition function for the liquid alkali metal halides:

$$
f = \left\{ \frac{e^{(E_s/2RT)(V/V_s)^{1/3}}}{(1 - e^{-\theta/T})^3} \left[1 + n\left(\frac{V - V_s}{V_s}\right) \exp\left(-\frac{aE_s V_s}{2(V - V_s)RT} \left(\frac{V}{V_s}\right)^{1/3} \right) \right] \right\}^{2N(V_s/V)}
$$

$$
\times \left\{ \frac{(2\pi m k T)^{3/2}}{h^3} \frac{eV}{N} \cdot \frac{8\pi^2 I k T}{h^2} \frac{1}{1 - e^{-h\nu/kT}} \right\}^{N \frac{V - V_s}{V}}
\tag{4.12}
$$

Some of the results for liquid KCl are shown in Table 4-8. Again, the model is successful. However, Carlson *et al.* [22] did not evaluate the heat capacities,

Table 4-8   Properties of Molten KCl (after Carlson *et al.* [22])

	T_m (°K)	V_m (cc/mole)	ΔS_m (e.u.)	T_b (°K)	V_b (cc/mole)	ΔS_b (e.u.)	T_c (°K)	V_c (cc/mole)	P_c (atm)
Calc.	1023	49.06	5.40	1684	71.20	21.63	3092	432	135.5
Obs.	1049	48.80	5.8	1680	—	23.1	—	—	—
%	−2.6	−0.53	−6.90	−0.24	—	−6.36	—	—	—

n: 6 a: 0.00300 I: 2.195×10^{-36} g cm^2
V_s: 41.57 cc/mole θ: 170 (°K)
ω: 305 cm^{-1} E_s: 54.15 Kcal/mole

transport properties, etc. Recent experimental work has indicated that appreciable concentrations of the dimer exist in the alkali halide vapor phase. Inclusion of such terms in the partition function should improve the results; in fact, Lu, Ree, Gerrard, and Eyring [30] obtained satisfactory results for a series of alkali halide molecules. The following partition function was used in their calculations:

$$f = \left\{ \left(\frac{e^{E_s/RT(V/V_s)^{1/3}}}{(1 - e^{-\theta/T})^3} \left[1 + n\left(\frac{V - V_s}{V_s}\right) \exp\left\{\frac{aE_s(V/V_s)^{1/3}V_s}{2RT(V - V_s)}\right\} \right] \right)^{2N(V_s/V)} \right.$$

$$\left. \times \left\{ \left(\frac{F_1 eV}{N}\right)^{N(V - V_s)/V} \left(1 + 2\frac{n_2}{n_1}\right)^{n_1} \left[\frac{KN}{eV}\left(\frac{n_1}{n_2} + 2\right)\right]^{n_2} \right\} \right\}$$

where

$$F_1 = \left(\frac{2\pi m kT}{h^2}\right)^{3/2} \cdot \frac{8\pi^2 I_1 kT}{h^2} \frac{1}{1 - e^{-h\nu/kT}} \tag{4.13}$$

and

$$K = \frac{F_2}{F_1^2} = e^{-\Delta H/RT} e^{\Delta S/R}$$

Here, K is the equilibrium constant between monomer and dimer, F_2 denotes the partition function of the dimer per unit volume, ΔH and ΔS are the heat of reaction and the entropy change, and n_1 and n_2 are the number of monomer and dimer molecules respectively; the remaining notation is as previously defined. Some of the results are shown in Table 4-9. The values of C_p at high temperatures are almost certainly too high and indicate needed improvement of the partition function.

Table 4-9   Calculated and Observed Thermodynamic Properties of the Molten Salts [30]

T (°K)	V (cc) Calc.	Obs.	P (atm) Calc.	Obs.	S (e.u.) Calc.	Obs.	C_p (cal/mole) Calc.	Obs.
				NaCl				
1074 (m.p)	37.42	37.56	0.00015	—	36.37	40.77	15.69	16.00
1290	40.38	40.62	0.00753	—	38.95	—	15.71	16.00
1783 (b.p)	46.99	48.80	1.00046	1.0000	44.10	—	24.46	—
(2219)	(152.59)	—	(141.96)	—	—	—	—	—
				KCl				
1043 (m.p)	48.77	48.80	0.00022	—	39.00	42.21	16.58	16.00
1200	51.96	51.91	0.00433	—	41.14	44.45	16.70	16.00
1680 (b.p)	63.06	64.47	1.00149	1.0000	47.38	—	26.12	—
(2144)	(151.13)	—	(211.43)	—	—	—	—	—

( ) indicate the calculated values

Now, let us examine the properties of molten HgX_2 from the theory. According to Jhon, Clemena, and Van Artsdalen, the partition functions of the molten HgX_2 is

$$f = \left\{ \frac{e^{E_s/RT}}{(1 - e^{-\theta/T})^5} \left[1 + n(x - 1) \exp - \frac{aE_s}{n(x - 1)RT} \right] \right.$$

$$\left. \times \prod_{i=1}^{4} \frac{1}{1 - e^{-hv_i/kT}} K^{\frac{1}{(2 + K^{-1/2})}} \right\}^{N \frac{V_s}{V}}$$

$$\cdot \left\{ \frac{(2\pi mkT)^{3/2}eV}{h^3 N} \cdot \frac{8\pi^2 IkT}{2h^2} \cdot \prod_{i=1}^{4} \frac{1}{1 - e^{-hv_i/kT}} \right\}^{\frac{N(V - V_s)}{V}} \qquad (4.14)$$

Here, the equilibrium constant can be expressed as

$$K = \frac{[HgX^+][HgX_3^-]}{[HgX_2]^2} = \frac{f_{HgX^+} f_{HgX_3^-}}{f_{HgX_2}^2} = \frac{n_{HgX^+} n_{HgX_3^-}}{n_{HgX_2}^2} = \frac{n_{HgX^+}^2}{n_{HgX_2}^2}$$

$$= e^{-\Delta H/RT} e^{\Delta S/R} \quad \text{since} \quad n_{HgX^+} = n_{HgX_3^-} \qquad (4.14a)$$

In formulating the partition function of the molten mercuric halides, the five degrees of freedom of an Einstein oscillator were assumed, on the basis that the HgX_2 salts have an unusually high entropy of fusion ($\Delta S_f = 7.5$ e.u. for $HgCl_2$, 8.4 e.u. for $HgBr_2$, and 8.6 e.u. for HgI_2) and that no solid-solid transitions have been observed. Therefore, probably no rotation occurs in the

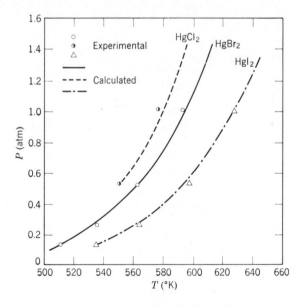

Figure 4-4 Vapor pressures of molten HgX_2. (after Jhon et al. [28])

solid state. The introduction of the equilibrium constant K in the partition function to take care of species such as HgX^+ and HgX_3^- follows the pattern developed earlier for the water partition function. (Water is discussed later.) For the gas-like partition function, we consider the monomeric molecules which are shown from density measurements [31] below 1000°C. Some of the results are shown in Table 4-10 and Figure 4-4. The results are quite

Table 4-10 Calculated and Observed Molar Volumes
of Molten Mercuric Halides [28]

T (°K)	V_{calc} (cc/mole)	V_{obs} (cc/mole)	$\Delta\%$
	$HgCl_2$		
550 (m.p)	62.21	62.21	0.00
559.2	62.68	62.60	0.13
577.0 (b.p)	63.50	63.34	0.25
590.0	64.07	63.90	0.27
610.0	64.94	64.77	0.26
	$HgBr_2$		
514.0 (m.p)	70.50	70.54	−0.06
563.0	72.69	72.80	−0.15
580.0	73.43	73.62	−0.26
592.0 (b.p)	73.97	74.20	−0.31
610.0	74.80	75.10	−0.40
	HgI_2		
530 (m.p)	86.90	86.90	0.00
534.95	87.18	—	—
550	87.96	88.00	0.04
570	88.94	89.11	0.19
597.35	90.29	—	—
627.15 (b.p)	91.81	—	—
640.0	92.50	—	—

satisfactory. In the calculation, the following parametric values were used:

	a	θ (°K)	E_s (cal/mole)	V_s (cc/mole)	ΔH (cal/mole)	ΔS (e.u.)
$HgCl_2$	0.04028	31.11	16530	56.79	—	—
$HgBr_2$	0.01783	26.26	17346	66.47	—	—
HgI_2	0.01764	19.98	17835	81.98	−9127	−31.76

4.7 LIQUID HYDROGEN

Significant structure theory has been applied with excellent results [32] to quantum liquids such as liquid hydrogen, deuterium, and hydrogen deuteride. These calculations provide another revealing test of the model since the partition function involves different forms of statistics and also the changes in the concentrations of ortho- and para-hydrogen cause slight changes in the thermodynamic properties. Because these liquids exist at extremely low temperatures, a Debye partition function was used for the solid-like molecules. At the temperatures of interest, the Bose-Einstein statistics is obeyed by hydrogen and deuterium and the Fermi-Dirac statistics by hydrogen deuteride. Henderson et al. [32] obtained the following partition function:

$$\ln f = N \frac{V_s}{V} \left\{ \frac{E_p}{RT} - \frac{9}{8} \left(\frac{\theta_D}{T} \right) - 9 \left(\frac{T}{\theta_D} \right)^3 \int_0^{\theta_D/T} U^2 \ln(1 - e^{-U})\, dU \right.$$

$$\left. + \ln \left(1 + Z \frac{V - V_s}{V} e^{-\alpha} \right) \right\}$$

$$+ N \frac{V - V_s}{V} \left\{ 1 - \ln y \pm \frac{y}{2^{5/2}} \right\} + N \ln f_r \tag{4.15}$$

where the top sign ($+$) applies to the Bose-Einstein gas and the bottom sign ($-$) to the Fermi-Dirac case; $\alpha = E_p V/RT(V - V_s)$, E_p is the potential energy of the lattice, θ_D is the Debye temperature of the solid-like lattice, $y = N/V(h^2/2\pi mkT)^{2/3}$, and f_r is the rotational partition function for the molecules.

If $\theta_r = h^2/8\pi^2 Ik$, then the rotational partition functions for ortho-hydrogen (o-H_2) and para-hydrogen (p-H_2), respectively, are given by the expressions:

$$f_r^o = 3 \sum_{n=1,3}^{\infty} (2n + 1)e^{-n(n+1)\theta_r/T}$$

$$f_r^p = \sum_{n=0,2}^{\infty} (2n + 1)e^{-n(n+1)\theta_r/T} \tag{4.16}$$

For ortho-deuterium (o-D_2) and para-deuterium (p-D_2), the corresponding expressions are:

$$f_r^o = 6 \sum_{n=0,2}^{\infty} (2n + 1)e^{-n(n+1)\theta_r/T}$$

$$f_r^p = 3 \sum_{n=1,3}^{\infty} (2n + 1)e^{-n(n+1)\theta_r/T} \tag{4.17}$$

The rotational partition function for mixtures of ortho and para molecules is then given by:

$$\ln f_r = \eta \ln f_r^o + (1 - \eta) \ln f_r^p \tag{4.18}$$

where η is the fraction of the molecules which are in the ortho state. For normal hydrogen, $\eta = \frac{3}{4}$, while for normal deuterium $\eta = \frac{2}{3}$. Finally, for hydrogen deuteride (H-D),

$$f_r = 6 \sum_{n=1,\,2}^{\infty} (2n + 1)e^{-n(n+1)\theta_r/T} \tag{4.19}$$

The values of the parameters used are given in Table 4-11. V_s, θ_p, and I are from experimental data.

Table 4-11   Observed and Calculated Values of the Parameters [32]

	p-H$_2$	n-H$_2$	H-D	o-D$_2$	n-D$_2$
V_s (cc/mole)	23.34	23.25	21.84	20.58	20.48
θ_D (°K)	91	91	90	89	89
E_p (cal/mole)	384.9	386.3	435.7	475.2	476.8
I (g cm$^2 \times 10^{-41}$)	4.67	4.67	6.21	9.31	9.31
θ_r (°K)	85.4	85.4	64.2	42.8	42.8
$a \times 10^2$	0.547	0.554	0.565	0.640	0.647

Table 4-12 lists some calculated and observed properties at the melting, boiling, and critical points. The vapor pressure versus temperature for several species are shown in Figure 4-5. It is evident that a straightforward application of the significant structure method provides good agreement with experiment. However, in this work, it was assumed that $\frac{2}{3}Z$ was the best value for n in the

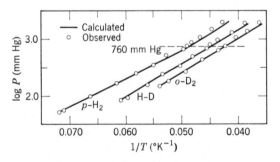

Figure 4-5   Vapor pressure of liquid para hydrogen, hydrogen deuteride, and ortho deuterium. (after Henderson *et al.* [32])

calculation of the critical constants. In this connection, a readily visualized model suffices to explain the isotopic pressure differences of various liquids. According to the findings of Grosh, Jhon, Ree, and Eyring [33], the vapor pressure differences are primarily determined by E_s, the heat of sublimation, but secondarily by the molecular structure parameters such as the moment of inertia I and the mass of isotopic molecules m. In the particular case of the H_2 and D_2 pair, the effect of I on the gas-like partition function is compensated for by its solid counterpart, since the rotational terms appear in both solid-like and gas-like partition functions. However, the E_s difference between the H_2 and D_2 is comparatively large (90 cal/mole); because $E_{sD} > E_{sH}$, naturally $P_{H_2} > P_{D_2}$.

Table 4-12 Calculated and Observed Thermodynamic Properties of Liquid Hydrogen [32]

	p-H$_2$	n-H$_2$	H-D	o-D$_2$	n-D$_2$	
T_m (°K)	(13.84)	(13.94)	(16.60)	(18.63)	(18.73)	calc
	13.84	13.94	16.60	18.63	18.73	obs
P_m (atm)	0.07388	0.07589	0.1236	0.1706	0.1724	calc
	0.06942	0.07085	0.1221	0.1678	0.1692	obs
V_m (cm³ mole⁻¹)	26.213	26.093	24.491	23.262	23.155	calc
	26.176	26.108	24.487	—	23.162	obs
ΔS_m (cal mole⁻¹ deg⁻¹)	1.932	1.936	2.048	2.198	2.210	calc
	2.028	—	—	2.526	—	obs
T_b (°K)	20.58	20.70	22.29	23.65	23.75	calc
	20.261	20.365	22.14	23.59	23.67	obs
V_b (cm³ mole⁻¹)	28.829	28.692	26.525	24.955	24.830	calc
	28.482	28.393	—	—	—	obs
ΔS_b (cal mole⁻¹ deg⁻¹)	10.553	10.564	11.868	12.741	12.737	calc
	10.602	—	—	12.459	—	obs
T_c (°K)	35.9	36.2	37.6	39.4	39.7	calc
	32.994	33.24	35.908	38.262	38.24	obs
P_c (atm)	13.6	13.8	15.5	17.1	17.3	calc
	12.770	12.797	14.645	16.282	16.421	obs
V_c (cm³ mole⁻¹)	77.7	77.3	71.5	68.3	68.0	calc
	65.5	—	62.8	60.3	—	obs

4.8 DENSE LIQUIDS AND GASES

The theory has also been successfully applied to dense liquids and gases [34]. Originally, in using an Einstein partition function for the solid-like degrees of freedom, it was assumed that $l\theta \gg T$, where l is the vibrational quantum number. The partition function for a one-dimensional harmonic oscillator terminating at the lth level is

$$\sum_{i=0}^{l-1} e^{i\theta/T} = \frac{1 - e^{-l\theta/T}}{1 - e^{-\theta/T}} \tag{4.20}$$

Here, l is determined by the relation $l\theta \simeq E_s/3R$. The usual Einstein partition function is obtained when $l\theta \geq T$. However, this relation is faulty for substances with low values of E_s (i.e., low molecular weight substances such as argon). For high energy levels, the one-dimensional Einstein oscillator is better described as a one-dimensional gas with translational degree of freedom, whose partition function is given by

$$e^{-l\theta/T} \frac{(2\pi mkT)^{1/2}}{h} v_f^{1/3} \tag{4.21}$$

where v_f is the molecular free volume in the solid and is represented by $v_f = [(V_s/N)^{1/3} - (b/4N)^{1/3}]^3$, b being the van der Waals constant, and $b/4N$ the net molecular volume. Accordingly, when $l\theta \gg T$ we replace the one-dimensional Einstein partition function $(1 - e^{-\theta/T})^{-1}$, by the expression,

$$\frac{1 - e^{-l\theta/T}}{1 - e^{-\theta/T}} + e^{-l\theta/T} \cdot \frac{(2\pi mkT)^{1/2}}{h} \cdot v_f^{1/3} \tag{4.22}$$

The partition function f_s in Eq. (3.7) thus becomes

$$f_s = e^{E_s/RT} \left\{ \frac{1 - e^{-l\theta/T}}{1 - e^{-\theta/T}} + e^{-l\theta/T} \cdot \frac{(2\pi mkT)^{1/2}}{h} \left[\left(\frac{V_s}{N}\right)^{1/3} - \left(\frac{b}{4N}\right)^{1/3} \right] \right\}^3$$

$$\times \left(1 + n \frac{V - V_s}{V_s} \exp \left[\frac{-aE_s V_s}{(V - V_s)RT} \right] \right) \tag{4.23}$$

where f_g is given by the usual partition function of three translational degrees of freedom. For nitrogen, the following partition function is used:

$$f = \frac{8\pi^2 IkT}{2h^2} (1 - e^{-h\nu/kT})^{-1} (f_s)^{N V_s/V} (f_g)^{\frac{N(V - V_s)}{V}} \tag{4.24}$$

This partition function for liquid nitrogen involves the rotational and vibrational partition function in both f_s and f_g.

The equation of state obtained from Eq. (3.2) modified by means of Eq. (4.23) for inert gases is

$$P = \frac{RT}{V_s x^2} \left[-\sigma_1 - \sigma_2 + x \cdot \left(\frac{\partial \sigma_2}{\partial x}\right)_T + x - 1 + \gamma + \ln x \right] \tag{4.25}$$

where

$$x = \frac{V}{V_s}$$

and

$$\sigma_1 = \frac{E_s}{RT} + 3 \ln \left\{ \frac{1 - e^{-l\theta/T}}{1 - e^{-\theta/T}} + e^{-l\theta/T} \cdot \frac{(2\pi mkT)^{1/2}}{h} \cdot v_f^{1/3} \right\}$$

$$\sigma_2 = \ln \left(1 + n(x - 1)e^{-aE_s/RT(x-1)} \right)$$

$$\gamma = \ln \left\{ \frac{(2\pi mkT)^{3/2}}{h^3} \cdot \frac{eV_s}{N} \right\} \qquad (4.26)$$

In addition to these modifications, the pressure effect on V_s must be considered. Thus,

$$V_{sp} = V_s(1 - \beta\Delta P) \qquad (4.27)$$

where V_{sp} indicates the solid volume under pressure, β is the solid compressibility, ΔP is the excess pressure above the vapor pressure at the melting point, and V_s has already been defined. Since β is rather small, 10^{-5} to 10^{-7} atm^{-1}, V_{sp} reduces to V_s if ΔP is small. For argon, the pressure effect is negligible if $\Delta P < 500$ atm. From Eq. (4.25), the second virial coefficient $B(T)$ may be determined:

$$B(T) = V_s(\gamma - \sigma_1 - \ln n) \qquad (4.28)$$

The calculated results are compared with experiment. Some results for nitrogen are shown in Figure 4-6. Figure 4-7 shows the compressibility factor,

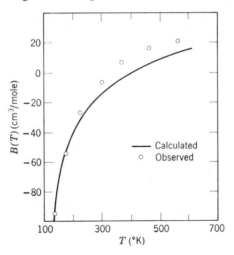

Figure 4-6 The second virial coefficients of nitrogen versus temperature. (After Ree, Ree, and Eyring [34])

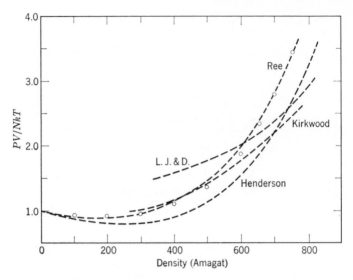

Figure 4-7 Compressibility factor of argon at 0°C. The points represent the experimental values. (after Ree, Ree, and Eyring [34])

PV/RT, as a function of density for argon at 0°C, and the results are compared with those of Wentorf *et al.* [35], Henderson [36], and Kirkwood [37]. The significant structure method agrees with experiment better than the other methods.

The values of the individual virial coefficients predicted by the significant structure method are also in fairly good agreement with experiment. The success of the significant structure theory throughout the entire gas and liquid region is highly satisfactory.

4.9 BINARY MIXTURES

The study of binary mixtures is one of the exciting problems in modern physical chemistry. Many qualitative and quantitative theories have been proposed to predict the properties of mixtures from a knowledge of those of the pure components. The significant structure approach works very well when applied to various classes of liquids; the following binary mixtures have been successfully treated: $CCl_4 + C_6H_{12}$ [19], $CCl_4 + C_6H_6$ [38], $C_6H_{12} + C_6H_6$ [38], $A + N_2$ [39], $A + O_2$ [39], $O_2 + N_2$ [39], $C_6H_6—C_2H_4Cl_2$ [40], and alkali halide mixtures [41].

According to the theory, there are three significant structures in a liquid: (1) molecules with solid-like degrees of freedom, (2) positional degeneracy in the solid-like structure, (3) molecules with gas-like degrees of freedom. We can extend the theory directly to binary mixtures, except that we must consider the concentration dependence of the parameters. Thus, the following assumptions were made: (a) nonrandom mixing is negligible; (b) the same characteristic temperatures of vibration, θ, are retained for mixtures that were used for the pure substances; (c) molecules of both components continue to possess their gas-like translational degrees of freedom; (d) the degeneracy term has the same form as for a pure liquid; (e) the parameters E_s, V_s, n, and a may be taken as suitable averages of the parameters for the pure components. With these assumptions, the partition function for a mixture takes the following form:

$$f_{\text{mixture}} = \frac{(N_1 + N_2)!}{N_1! N_2!} \left[f_{s_1}^{N_1} f_{s_2}^{N_2} \right]^{V_s/V} \cdot \left[f_{\text{deg}} \, e^{E_s/RT} \right]^{(N_1 + N_2) V_s/V} \cdot \left[f_{g_1}^{N_1} f_{g_2}^{N_2} \right]^{(V - V_s)/V}$$

$$(4.29)$$

where

$$f_{s_1} = \frac{1}{(1 - e^{-\theta_1/T})^3} \cdot f_{\text{rot 1}}^{(s)} f_{\text{vib 1}}$$

$$(4.30a)$$

Here, we assume that the solid-like molecule rotates:

$$f_{g_1} = \frac{(2\pi m k T)^{3/2}}{h^3} f_{\text{rot 1}}^{(g)} f_{\text{vib 1}} \cdot \frac{eV}{N_1 + N_2}$$

$$(4.30b)$$

with similar equations for component 2, and

$$f_{\text{deg}} = 1 + n \left(\frac{V - V_s}{V_s} \right) \exp \left(\frac{-a E_s V_s}{(V - V_s)RT} \right)$$

$$(4.30c)$$

$$E_s = X_1^2 E_{s_1} + X_2^2 E_{s_2} + 2(1 + \delta_E) X_1 X_2 \sqrt{E_{s_1} E_{s_2}}$$

$$(4.30d)$$

$$V_s = X_1 V_{s_1} + X_2 V_{s_2} + X_1 X_2 \delta_V \sqrt{V_{s_1} V_{s_2}}$$

$$(4.30e)$$

$$n = X_1 n_1 + X_2 n_2$$

$$(4.30f)$$

$$a = X_1 a_1 + X_2 a_2$$

$$(4.30g)$$

The quantities δ_E and δ_V are the only parameters in the mixture partition function which were not evaluated from the pure liquids. The quantities, δ_E and δ_V are the correction parameters used in the cross terms and in the terms containing higher powers of the concentration. In general, these values are very small, and reasonable results are obtained when they are taken to be zero.

At constant temperatures, the equilibrium condition imposes the set of relationships:

$$P^L = P^G \qquad -\left(\frac{\partial A}{\partial V}\right)^L_{T, N_1, N_2} = -\left(\frac{\partial A}{\partial V}\right)^G_{T, N_1, N_2}$$

$$\mu_1^L = \mu_1^G \qquad \left(\frac{\partial A}{\partial N_1}\right)^L_{T, V, N_2} = \left(\frac{\partial A}{\partial N_1}\right)^G_{T, V, N_2} \qquad (4.31)$$

$$\mu_2^L = \mu_2^G \qquad \left(\frac{\partial A}{\partial N_2}\right)^L_{T, V, N_1} = \left(\frac{\partial A}{\partial N_2}\right)^G_{T, V, N_1}$$

where μ_i is the chemical potential of the ith component. To calculate the thermodynamic properties and the excess properties of a mixture, we must first find two points on the free-energy surface which make the three conditions given by Eq. (4.31) hold true. Some of the excess properties are summarized as follows:

$$G^E = G - (X_1 G_1^0 + X_2 G_2^0) - RT(X_1 \ln X_1 + X_2 \ln X_2)$$

$$S^E = S - (X_1 S_1^0 + X_2 S_2^0) + R(X_1 \ln X_1 + X_2 \ln X_2)$$

$$E^M = E - (X_1 E_1^0 + X_2 E_2^0)$$

$$A^E = A - (X_1 A_1^0 + X_2 A_2^0) - RT(X_1 \ln X_1 + X_2 \ln X_2) \qquad (4.32)$$

$$V^E = V - (X_1 V_1^0 + X_2 V_2^0)$$

$$C_p^E = C_p - (X_1 C_p^0 + X_2 C_p^0)$$

Some of the results for the excess quantities are shown in Tables 4-13, 4-14, and 4-15 and in Figures 4-8 and 9. Agreement between experiment and theory is satisfactory.

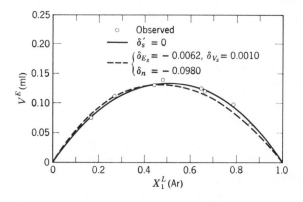

Figure 4-8   Excess volumes of the $Ar + O_2$ system at $83.82°K$ and constant pressure. (after Miner and Eyring [39])

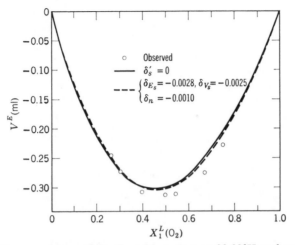

Figure 4-9 Excess volumes of the $O_2 + N_2$ system at 83.82°K and constant pressure. (after Miner and Eyring [39])

Table 4-13 Parameters and Constants for Single Components

Component	θ	V_s	E_s	n	a	I (gr cm²)	ν (sec⁻¹)
Ar	60.00	24.98	1888.6	10.8	5.34×10^{-3}	—	—
N_2	55.94	29.31	1529.9	12.9	3.43×10^{-3}	13.94×10^{-40}	7.075×10^{13}

Table 4-14 Excess Properties of the Ar + N_2 System at 83.82°K and Constant Pressure

X_1^L (Ar)	G^E (cal/mole) Calc.		Obs.	H^E (cal/mole) Calc.		Obs.	TS^E (cal/mole) Calc.		C_p^E (cal/mole°K) Calc.	
0.1156	2.00	3.23	—	3.00	4.73	5.6	1.02	1.52	−0.132	−0.125
0.2243	3.45	5.55	—	5.29	8.20	9.7	1.86	2.66	−0.242	−0.232
0.3482	4.53	7.30	—	7.21	10.99	11.6	2.68	3.70	−0.346	−0.334
0.4315	4.94	7.93	—	8.06	12.13	12.2	3.12	4.20	−0.400	−0.389
0.5000	5.05	8.12	8.2	8.44	12.59	12.1	3.39	4.47	−0.432	−0.422
0.5420	5.03	8.08	—	8.54	12.66	12.0	3.50	4.56	−0.445	−0.437
0.6110	4.84	7.77	—	8.44	12.38	11.4	3.59	4.59	−0.454	−0.448
0.7229	4.10	6.59	—	7.52	10.84	9.8	3.39	4.23	−0.429	−0.426
0.8151	3.10	4.98	—	5.95	8.46	7.5	2.83	3.45	−0.358	−0.359
0.9387	1.10	1.87	—	2.46	3.43	2.7	1.28	1.52	−0.161	−0.163

Column 1: $\delta_v = 0$, $\delta_E = 0$; Column 2: $\delta_E = -0.0043$, $\delta_v = -0.009$;
Column 3; observed

Table 4-15   Some Additional Properties of the Ar $+$ N$_2$ System at 83.82°K

X_1^l (Ar)	V_E (ml/mole) Calc.		Obs.	$\alpha \times 10^3$ (deg^{-1})		$\beta \times 10^5$ (atm^{-1})	
0.1156	−0.069	−0.084	−0.075	5.952	5.967	37.86	37.99
0.2243	−0.112	−0.138	−0.124	5.740	5.764	34.59	34.79
0.3482	−0.137	−0.174	−0.163	5.528	5.555	31.33	31.54
0.4315	−0.141	−0.182	−0.175	5.402	5.429	29.39	29.59
0.500	−0.138	−0.181	−0.179	5.310	5.236	27.93	28.11
0.5420	−0.133	−0.177	−0.177	5.260	5.284	27.09	27.25
0.6110	−0.122	−0.165	−0.168	5.185	5.207	25.80	25.94
0.7229	−0.095	−0.132	−0.145	5.093	5.108	23.94	24.03
0.8151	−0.066	−0.094	−0.111	5.048	5.048	22.61	22.67
0.9387	−0.022	−0.033	−0.042	5.046	5.048	21.13	21.15

Column 1; $\delta_E = 0$, $\delta_v = 0$;   Column 2; $\delta_E = 0.0043$, $\delta_v = -0.0090$

4.10   WATER AND IONIC SOLUTIONS

An exciting problem is the theory of water structure. The successful application of the theory to water is the prelude to a general attack on the theory of ionic solutions. Water is the most common liquid on earth, but it is abnormal in many respects: its melting point, boiling point, heat of vaporization, and heat of fusion are all higher than would normally be expected from the hydride compounds of the other members of the oxygen family. Many theories have been proposed to explain the properties of water and to elucidate its structure. Unfortunately, these do not explain all its properties quantitatively. This is especially true for the abnormal physical properties, such as the maximum density at 4°C, and the peculiar decrease in viscosity with pressure up to about 1000 atmospheres.

Typical earlier models include the Bernal-Fowler postulate of quartz-like structure [42], Pople's, bent-hydrogen-bond model [43], Pauling's dodecahedral cage model [44], and Frank and Wen's flickering cluster model [45]. Recently, Nemethy and Scheraga [46] have applied a modified form of significant structure theory to the Frank and Wen flickering cluster model. They were successful in explaining the maximum density of H$_2$O and D$_2$O. Their calculated cluster size at 30°C is approximately equal to Pauling's. However, they did not write down a partition function from which all the properties were then calculated. The significant structure approach has also been applied to water by Eyring and Marchi [47], who assumed two solid-like species, a nonrotating hydrogen-bonded species and a rotating monomer, and calculated all thermodynamic variables with reasonably good results. Stevenson

[48] interprets his absorption spectroscopic studies for liquid water to show that the rotating monomer concentration is in much lower concentration than the Eyring-Marchi calculation indicates. Their model failed to yield the density maximum at 4°C.

To overcome these difficulties, Jhon, Grosh, Ree, and Eyring [49] proposed a new model in which water is visualized as containing at least two solid-like structures in equilibrium with each other and with the gas-like molecules. One of these structures is a cluster of about 46 molecules with a structure and density similar to ice I. The clusters are dispersed in the ice-III-like structure which is 20% more dense than ice I but is still hydrogen bonded. An equilibrium is established between the ice-I-like and ice-III-like structures so that when a cluster of ice-I-like molecules disappears, the molecules change their structure to 46 ice-III-like molecules. Introducing these concepts, we can write a partition function for water.

The partition function for solid-like degrees of freedom is given by

$$f_s^{V_s N/V} = (f_{s_I})^{n_I}(f_{s_{III}})^{n_{III}} \tag{4.33}$$

where f_{s_I} and $f_{s_{III}}$ are the partition function for the ice-I-like and ice-III-like molecules, and n_I and n_{III} are the numbers of molecules of the respective species. From the significant structure theory, Eq. (4.33) may be written as

$$f_s = \frac{\exp(E_s/RT)}{(1 - e^{-\theta/T})^6} \left[1 + n(x-1)\exp\left(-\frac{aE_s}{(x-1)RT}\right) \right]$$

$$\cdot \prod_{i=1}^{3} \frac{1}{1 - e^{-hv_i/kT}} (K)^{\frac{K}{(1+K)q}} \tag{4.34}$$

In deriving Eq. (4.34), we assumed that both ice-I-like and ice-III-like molecules do not rotate and that between them an equilibrium is established:

Ice-I-like molecules $\rightleftarrows$ Ice-III-like molecules

Here, the cluster of q ice-I-like molecules changes cooperatively to q ice-III-like molecules and the equilibrium constant is thus written as

$$K = [\text{Ice-I-like}]/[\text{Ice-III-like}] = \left(\frac{f_{s_I}}{f_{s_{III}}}\right)^q$$

$$= \left[\exp\left(-\frac{\Delta H}{RT}\right) \exp\left(\frac{\Delta S}{R}\right) \exp\left(-\frac{P\Delta V}{RT}\right) \right]^q \tag{4.35}$$

and

$$V_s = \frac{K}{1+K} V_{s_I} + \frac{1}{1+K} V_{s_{III}} = \frac{KV_{s_I} + V_{s_{III}}}{1+K} \tag{4.36}$$

For the gas-like partition function, the following form is used:

$$f_g = \frac{(2\pi m k T)^{3/2}}{h^3} \frac{eV}{N} \cdot \frac{8\pi^2 (8\pi^3 ABC)^{1/2}(kT)^{3/2}}{2h^3} \cdot \prod_{i=1}^{3} \frac{1}{1 - e^{-hv_i/kT}} \quad (4.37)$$

where all the symbols have been defined previously. This yields the following partition function for water:

$$f = \left\{ \frac{e^{E_s/RT}}{(1 - e^{-\theta/T})^6} [1 + n(x - 1)e^{-\frac{aE_s}{(x-1)RT}}](K)^{\frac{K}{(1+K)q}} \prod_{i=1}^{3} \frac{1}{1 - e^{-hv_i/kT}} \right\}^{NV_s/V}$$

$$\cdot \left\{ \frac{(2\pi m k T)^{3/2}}{h^3} \frac{eV}{N} \cdot \frac{8\pi^2 (8\pi^3 ABC)^{1/2}(kT)^{3/2}}{2h^3} \cdot \prod_{i=1}^{3} \frac{1}{1 - e^{-hv_i/kT}} \right\}^{\frac{N(V - V_s)}{V}}$$

$$(4.38)$$

Some of the calculated results are shown in Figure 4-10 and in Tables 4-16, 4-17, and 4-18. The theory successfully predicts the maximum density at 4°C, and a minimum in the heat capacity versus temperature curve, as is observed. The results for D_2O are similar to those for water, and may be found in the original paper.

Table 4-16   Parametric Values for Water

$E_s = 10760$ cal/mole	$a = 0.1653 \times 10^{-4}$
$V_{s_I} = 19.65$ cc/mole[a]	$\theta = 216.1°K$
$V_{s_{III}} = 17.65$ cc/mole	$\Delta H = -481.5$ cal/mole
$n = 11.94$	$\Delta S = -1.863$ e.u.
$q = 46$	

[a] The observed volume of ice at 0°C.

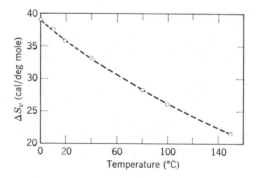

Figure 4-10 Entropy of vaporization vs. temperature. H_2O : O, experimental; ————, calculated. (after Jhon et al. [49])

Table 4-17 Molar Volumes and Vapor Pressures of Water

T (°K)	V_{calc} (cc)	V_{obs} (cc)	% error	P_{calc} (atm)	P_{obs} (atm)	% error
273.15 (m.p)	17.923	18.019	−0.53	0.006051	0.006030	−0.35
277.15	17.888	18.016	−0.71	0.008036	0.008007	−0.36
283.15	17.891	18.021	−0.72	0.01210	0.01218	−0.66
293.15	17.961	18.048	−0.48	0.02298	0.02307	−0.39
313.15	18.171	18.157	0.08	0.07234	0.07279	−0.62
353.15	18.673	18.538	0.73	0.4623	0.4672	−1.05
373.15 (b.p)	18.959	18.799	0.85	0.9852	1.0000	−1.48
423.15	19.800	19.641	0.81	4.549	4.698	−3.17

Table 4-18 Specific Heats of Water

T (°K)	C_p (obs) (cal/mole)	C_p (calc) (cal/mole)	% error
273.15 (m.p)	18.15	17.40	−4.13
283.15	18.06	15.84	−12.3
293.15	17.99	16.61	−7.67
313.15	17.98	17.39	−3.28
353.15	18.06	18.94	−0.11
373.15 (b.p)	18.14	18.33	1.05
423.15	—	19.04	—

Work is just beginning on the application of significant structure theory to ionic solutions. A partition function which is being considered is

$$
f_{\text{ionic solution}} = \left\{ \frac{\exp\left(\dfrac{E_{smx}}{RT}\right)}{(1 - e^{-\theta_1/T})^3(1 - e^{-\theta_2/T})^3} \right\}^{m'N} (K_1)^{\frac{K_1 m' n_1 N}{1+K_1}} (K_2)^{\frac{K_2 m' n_2 N}{1+K_2}}
$$

$$
\cdot \left\{ \frac{\exp\left(\dfrac{E_s}{RT}\right)}{(1 - e^{-\theta/T})^6} (K)^{\frac{K}{q(1+K)}} \right.
$$

$$
\left. \times \left(1 + n\frac{V - V_s}{V_s} e^{-\frac{aE_sV_s}{(V-V_s)RT}}\right) \prod_{i=1}^{3} \frac{1}{1 - e^{-hv_i/kT}} \right\}^{55.51NV_s/V}
$$

$$\cdot \left\{ \left(\frac{2\pi m k T}{h^2} \right)^{3/2} \cdot \frac{eV}{N} \right.$$

$$\left. \cdot \frac{8\pi^2 (8\pi^3 ABC)^{1/2} (kT)^{3/2}}{2h^3} \prod_{i=1}^{3} \frac{1}{1 - e^{-hv_i/kT}} \right\}^{55.51N \frac{V - V_s}{V}}$$

$$\cdot \exp \left\{ \frac{V}{8\pi} \left(\frac{1}{1 + a'\kappa} \left(\kappa^3 - \frac{1}{a'^3} \right) \right. \right.$$

$$\left. \left. + \frac{1}{a'^3} (1 + a'\kappa) - \frac{2}{a'^3} \ln(1 + a'\kappa) \right) \right\}$$

$$\cdot \frac{(55.51N + 2m'N)!}{(55.51N)! ((m'N)!)^2} \qquad (4.39)$$

where m', K_1, K_2, θ_1, θ_2, N_1, N_2, E_{smx}, a' and κ represent the molality of the salt, the equilibrium constant between water absorbed on a site on a cation and free water, the corresponding equilibrium constant for the anion, the Einstein characteristic temperature for the cation, and for the anion, the number of sites on an anion, and on a cation, the energy of sublimation of solute, the ionic radius and the Debye Hückel parameter. The other notation has been defined previously. The results so far obtained are encouraging.

4.11 PLASTIC CRYSTALS

A new development in the theory of states of matter is the recognition of a state which Timmermans calls a " plastic crystal." This state normally occurs between the melting point and the transition point of substances composed of molecules with a high degree of symmetry. Many of the properties of this state suggest that the plastic crystal state is closely related to the liquid state. Since the significant structure theory is able to predict quite accurately all the thermodynamic and many physical properties of a liquid, it seems natural that this approach should be applied to the plastic crystal state.

Zandler and Thomson [50], applied the significant structure approach to the plastic crystal state of CBr_4, with good success. The partition function for the plastic crystal of CBr_4 has the following form:

$$f = \left\{ \frac{e^{E_s/RT}}{(1 - e^{-\theta/T})^6} \left[1 + Z \frac{V - V_s}{V} e^{-\frac{aE_sV_s}{(V - V_s)RT}} \right] \prod_{i=1}^{9} \frac{1}{1 - e^{-hv_i/kT}} \right\}^{N \frac{V_s}{V}}$$

$$\cdot \left\{ \frac{(2\pi m k T)^{3/2}}{h^3} \frac{eV}{N} \frac{8\pi^2 (8\pi^3 ABC)^{1/2} (kT)^{3/2}}{12h^3} \prod_{i=1}^{9} \frac{1}{1 - e^{-hv_i/kT}} \right\}^{N \frac{V - V_s}{V}} \qquad (4.40)$$

Some of the results are listed in Table 4-19 and are quite impressive.

Table 4-19 The Calculated and Observed Properties of Carbon Tetrabromide in the Plastic Crystal State

	T °K	P mm Hg	V cc/mole	S cal/°mole	H cal/mole	H_{sub} cal/mole	$\alpha \times 10^4$ °K^{-1}	$\beta \times 10^6$ °K^{-1}	C_v cal/°mole	C_p cal/°mole
Obs.	320.00 (T_{tr})	3.14	102.91	—	—	11800	—	—	25.88	—
Calc.		(3.14)	(102.91)	59.59	−6941	12295	6.78	46.4	26.05	33.96
Obs.	325.00	4.20	103.18	—	—	11800	6.10	38.0	25.81	33.6
Calc.		4.23	103.24	60.11	−6772	12238	6.43	45.3	26.11	33.54
Obs.	330.00	5.62	103.50	—	—	11800	5.77	38.9	25.78	33.11
Calc.		5.63	103.57	60.62	−6605	12182	6.32	45.7	26.17	33.41
Obs.	335.00	7.33	103.80	—	—	11800	5.82	—	25.74	33.03
Calc.		7.42	103.90	61.13	−6438	12127	6.30	46.7	26.22	33.40
Obs.	340.00	9.57	104.14	—	—	—	6.02	—	25.71	33.09
Calc.		9.70	104.23	61.62	−6271	12071	6.35	48.2	26.26	33.45
Obs.	345.00	12.3	104.46	—	—	—	—	—	25.68	33.25
Calc.		12.56	104.56	62.11	−6103	12016	6.42	49.9	26.33	33.54
Obs.	350.00	15.9	104.77	—	—	—	—	—	25.65	33.38
Calc.		16.12	104.90	62.60	−5935	11961	6.51	51.8	26.38	33.65
Obs.	355.00	20.3	105.10	—	—	—	—	—	25.62	33.56
Calc.		20.54	105.25	63.07	−5767	11905	6.62	53.9	26.43	33.77
Obs.	360.00	25.7	—	—	—	—	—	—	25.60	33.90
Calc.		25.95	105.60	63.55	−5598	11849	6.73	56.2	26.47	33.90
Obs.	363.25 (T_{tp})	30.1	—	—	—	—	—	—		
Calc.		(30.1)	105.83	63.85	−5487	11813	6.81	57.7	26.50	33.99

4.12 THE SHOCK WAVE COMPRESSION OF ARGON

The shock compression of argon is reported by Van Thiel and Alder [51] for two initial states of 86°K and 2 bar, and 148.2°K and 70 bar. Very recently, the significant structure theory of liquids has been applied to the shock compression of argon [52]. The following corrections for the pressure effects in the partition function are introduced. The Tait equation [53], which is expressed as $V'_s = V_s - c \ln (B + P)/P$, for the molar volume of the solid V_s, the Gruneisen constant [54] which is defined by $\gamma = -(d \ln \theta/d \ln V_s)$, for the Einstein temperature θ, and the Lennard-Jones potential [52] for calculating E_s is used. The Hugoniot adiabatics can be obtained from the Rankine-Hugoniot jump conditions [53] and takes the form:

$$E - E_0 = \tfrac{1}{2}(P + P_0)(V_0 - V) \qquad (4.41)$$

where E_0, P_0, and V_0 represent the initial internal energy, the initial pressure, and initial volume, respectively. To obtain Eq. (4.41) one makes use of the equations for conservation of mass, energy, and momentum in a shock wave. Using the numerical iteration method, the calculated relation between V and P which represent the Hugoniot adiabatic are calculated using significant structure theory and are compared with experiment [52] in Figures 4-11 and 4-12. The agreement is highly satisfactory. The success of the model at these very high pressures is one more evidence of the general applicability of significant structure theory.

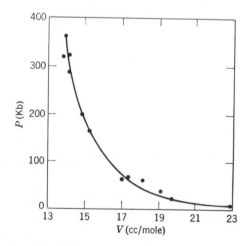

Figure 4-11 Shock compression of argon ($P_0 = 2$ bar, $\rho_0 = 1.405$ g/cc). ——— Calculated. · · · Experimental. (after Lin *et al.* [52])

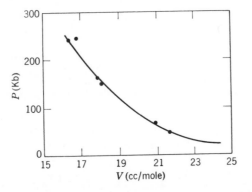

Figure 4-12 Shock compression of argon ($P_0 = 70$ bar, $\rho_0 = 0.919$ g/cc). ——— Calculated. $\cdots$ Experimental. (after Lin *et al.* [52])

REFERENCES

[1] J. Grosh, M. S. Jhon, T. Ree, and H. Eyring, *Proc. Natl. Acad. Sci. (U.S.)*, **57**, 1566 (1967).

[2] E. J. Fuller, T. Ree, and H. Eyring, *Proc. Natl. Acad. Sci. (U.S.)*, **45**, 1594 (1959).

[3] K. C. Kim, W. C. Lu, T. Ree, and H. Eyring, *Proc. Natl. Acad. Sci. (U.S.)*, **57**, 861 (1967).

[4] T. R. Thomson, H. Eyring, and T. Ree, *Proc. Natl. Acad. Sci. (U.S.)*, **46**, 336 (1960).

[5] T. R. Thomson, H. Eyring, and T. Ree, *J. Phys. Chem.*, **67**, 2701 (1963).

[6] H. Paik and S. Chang, *J. Korean Chem. Soc.*, **7**, 179 (1963).

[7] (a) M. S. Jhon, J. Grosh, T. Ree, and H. Eyring, *J. Phys. Chem.*, **70**, 1591 (1966).
(b) H. Lee, M. S. Jhon, and S. Chang, *J. Korean Chem. Soc.*, **8**, 1791 (1964).

[8] M. S. Jhon, J. Grosh, and H. Eyring, *J. Phys. Chem.*, **71**, 2253 (1967).

[9] J. Grosh, M. S. Jhon, T. Ree, and H. Eyring, *Proc. Natl. Acad. Sci. (U.S.)*, **54**, 1004 (1965).

[10] M. E. Zandler, J. A. Watson, and H. Eyring, *J. Phys. Chem.*, **72**, 2730 (1968).

[11] R. Schmidt, M. S. Jhon, and H. Eyring, *Proc. Natl. Acad. Sci. (U.S.)*, **60**, 387 (1968).

[12] E. J. Fuller, "Significant Liquid Structure," Ph.D. Thesis, Department of Chemistry, University of Utah, 1960.

[13] S. Chang, H. Pak, W. Paik, S. H. Park, M. S. Jhon, and W. S. Ahn, *J. Korean Chem. Soc.*, **8**, 33 (1964).

[14] H. Lee and S. Chang, *J. Korean Chem. Soc.*, **9**, 211 (1965).

[15] M. S. Jhon, J. Grosh, and H. Eyring, unpublished work.

[16] W. K. Paik and S. Chang, *J. Korean Chem. Soc.*, **8**, 29 (1964).

[17] W. S. Ahn and S. Chang, *J. Korean Chem. Soc.*, **8**, 125 (1964).

[18] M. S. Jhon, J. Grosh, T. Ree, and H. Eyring, *Proc. Natl. Acad. Sci. (U.S.)*, **54**, 1419 (1965).

[19] K. Liang, H. Eyring, and R. P. Marchi, *Proc. Natl. Acad. Sci. (U.S.)*, **52**, 1107 (1964).

[20] D. R. McLaughlin and H. Eyring, *Proc. Natl. Acad. Sci. (U.S.)*, **55**, 1031 (1966).

[21] C. M. Carlson, H. Eyring, and T. Ree, *Proc. Natl. Acad. Sci. (U.S.)*, **46**, 649 (1960).

[22] C. M. Carlson, H. Eyring, and T. Ree, *Proc. Natl. Acad. Sci. (U.S.)*, **46**, 333 (1960).

[23] G. E. Blomgren, *Ann. N. Y. Acad. Sci.*, **79**, 781 (1960).

[24] D. A. McQuarrie, *J. Phys. Chem.*, **66**, 1508 (1962).

[25] J. E. Lennard-Jones and A. Devonshire, *Proc. Roy. Soc.*, **169A**, 317 (1937).

[26] J. Kirkwood and E. Monroe, *J. Chem. Phys.*, **9**, 514 (1941).
J. Kirkwood and E. Boggs, *ibid.*, **10**, 394 (1942).
N. N. Bogolyubov, *Problemy Dinamicheskoi Teorii Statisticheskoi Fizike*, Gostekhizdct, Moscow, 1946.
M. Born and H. Green, *Proc. Roy. Soc.*, **188A**, 10 (1946); *ibid.*, **189A**, 255 (1947).

[27] G. Janz and J. McIntire, *Ann. N. Y. Acad. Sci.*, **79**, 790 (1960).

[28] M. S. Jhon, G. Clemena, and E. R. Van Artsdalen, *J. Phys. Chem.*, **72**, 4155 (1968).

[29] W. J. Hamer, *The Structure of Electrolytic Solutions*, Wiley, New York, 1959.

[30] W. C. Lu, T. Ree, V. G. Gerrard, and H. Eyring; *J. Chem. Phys.*, **49** 797 (1968)

[31] H. Braune and S. Knoke, *Z. Physik. Chem. (Leipzig)*, **A152**, 409 (1931).

[32] D. Henderson, H. Eyring, and D. Felix, *J. Phys. Chem.*, **66**, 1128 (1962).

[33] J. Grosh, M. S. Jhon, T. Ree, and H. Eyring, *Proc. Natl. Acad. Sci. (U.S.)*, **58**, 2196 (1967).

[34] T. S. Ree, T. Ree, and H. Eyring, *Proc. Natl. Acad. Sci. (U.S.)*, **48**, 501 (1962).

[35] R. H. Wentorf, R. J. Buehler, J. O. Hirschfelder, and C. F. Curtiss, *J. Chem. Phys.*, **18**, 1484 (1950).

[36] D. Henderson, *J. Chem. Phys.*, **37**, 631 (1962).

[37] J. G. Kirkwood, V. A. Lewinson, and B. J. Alder, *J. Chem. Phys.*, **20**, 929 (1952).

[38] S. M. Ma and H. Eyring, *J. Chem. Phys.*, **42**, 1920 (1965).

[39] B. A. Miner and H. Eyring, *Proc. Natl. Acad. Sci. (U.S.)*, **53**, 1227 (1965).

[40] W. S. Ahn, H. Pak, and S. Chang, *J. Korean Chem. Soc.*, **9**, 215 (1965).

[41] R. Vilen and C. Misdolea, *J. Chem. Phys.*, **45**, 3414 (1966).

[42] J. D. Bernal and R. H. Fowler, *J. Chem. Phys.*, **1**, 515 (1933).

[43] J. A. Pople, *Proc. Roy. Soc. (London)*, **A205**, 163 (1951).

[44] L. Pauling, in *Hydrogen Bonding*, edited by L. Hadzi, Pergamon, London, 1959, p. 1.

[45] H. S. Frank and W. Y. Wen, *Discussions Faraday Soc.*, **24**, 133 (1957).

[46] G. Nemethy and H. Scheraga, *J. Chem. Phys.*, **36**, 3382, 3401 (1962); **41**, 680 (1964).

[47] R. P. Marchi and H. Eyring, *J. Phys. Chem.*, **68**, 221 (1964).

[48] D. P. Stevenson, *J. Phys. Chem.*, **69**, 2145 (1965).

[49] M. S. Jhon, J. Grosh, T. Ree, and H. Eyring, *J. Chem. Phys.*, **44**, 1465 (1966).

[50] M. E. Zandler and T. R. Thomson, *Solid State Communications*, **4**, 219 (1966).

[51] M. Van Thiel and B. J. Alder, *J. Chem. Phys.*, **44**, 1056 (1966).

[52] S. H. Lin, D. Tweed, and H. Eyring, *Proc. Natl. Acad. Sci. (U.S.)*, in press.

[53] J. O. Hirschfelder, C. F. Curtiss, and R. B. Bird, *Molecular Theory of Gases and Liquid*, Wiley, New York, 1954.

[54] J. C. Slater, *Introduction to Chemical Physics*, McGraw-Hill, New York, 1938.

chapter 5

TRANSPORT PROPERTIES OF THE LIQUID

5.1 INTRODUCTION

In Chapters 3 and 4, significant structure theory is shown to yield good agreement between experimental thermodynamic properties and those calculated from the Helmholtz free energy, A, as a function of V and T. According to this model, a fraction V_s/V of the molecules shows solid-like behavior and the remaining fraction $(V - V_s)/V$ is gas-like. Molecules in the gas phase are assumed to be in approximately one-to-one correspondence with liquid vacancies that execute gas-like motions. We may assume that this model is in fact correct or that at least it differs very little from the true model as measured by the free energy.

Because of the successful treatment of thermodynamic properties, it is interesting to see how the significant structure theory explains transport phenomena. First, we describe the theory of viscosity and diffusion from significant structure theory and then we develop a theory of thermal conductivity.

5.2 THEORY OF VISCOSITY AND DIFFUSION

There are several theoretical approaches to the study of transport phenomena of liquids and dense gases. One was developed by Kirkwood [1]. Starting from Liouville's equation, he obtained exact expressions for the flux vectors in terms of nonequilibrium radial distribution functions. This formal approach is very attractive conceptually but, unfortunately, it is still not developed for practical work. Another procedure was introduced by Enskog [2]. Although the Enskog theory is derived for rigid-sphere systems, it has been applied to real dense gases in excellent agreement with experiment. Significant structure theory as developed by Eyring, Ree, and co-workers [3–5] applies absolute reaction rate theory to various transport phenomena. Since the flow of a liquid is a rate process, insofar as it takes place at a definite velocity under given conditions, it is natural to apply the theory of absolute reaction rates to the problems of transport. By definition, the viscosity η satisfies the following relation:

$$\eta = \frac{f}{\dot{s}} \qquad (5.1)$$

where f is the shear stress and $\dot{s}$ is the rate of shear. For simple liquids, a fraction X_s of a shear surface is covered by solid-like molecules and the remaining fraction X_g is covered by gas-like molecules. Then, the viscosity η, defined as the ratio of shear stress f to the shear rate $\dot{s}$, is given by

$$\eta = \frac{f}{\dot{s}} = (X_s f_s + X_g f_g)/\dot{s} = X_s \eta_s + X_g \eta_g = \frac{V_s}{V} \eta_s + \frac{V - V_s}{V} \eta_g \qquad (5.2)$$

Here f_s and f_g are the shear stresses acting on the solid-like and the gas-like molecules, respectively, and η_s and η_g are the viscosities contributed by the two types of molecules. In more complicated systems, there are additional kinds of flowing structures. Following Eyring's earlier procedure [6], we now calculate the viscosity of solid-like molecules. We write

$$\eta_s = \frac{f_s}{\dot{s}} = f_s \bigg/ \left[\sum_i \frac{\lambda \cos \theta_i}{\lambda_1} k_i \exp \left(\frac{\lambda_2 \lambda_3 \lambda f_s \cos \theta_i}{2kT} \right) \right] \qquad (5.3)$$

where $\dot{s}$ is the rate of shear corresponding to the velocity with which one molecular layer slips over the other divided by λ_1, the distance between layers; $\lambda_2 \lambda_3$ is the area occupied by a molecule on which the shear stress f_s is acting (see Figure 5-1); k_i is the frequency of jumping into the ith neighboring empty lattice site when $f = 0$ and is assumed to be equal to k' for every position i. When f is not zero, the work done in forcing the molecule forward is

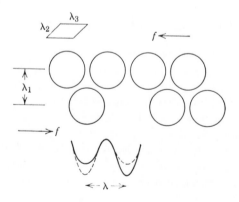

Figure 5-1 A schematic diagram of viscous shear. The area of a molecule in the shear plane is seen to be $\lambda_2 \lambda_3$ and the distance between flowing layers is λ_1.

the force $f_s \lambda_2 \lambda_3$ multiplied by the component of the distance to the top of the barrier $(\lambda \cos \theta_i)/2$, θ_i being the angle between the direction of stress and the direction of jumping. Thus, the rate of jumping into the ith site is $k_i \exp (f_s \lambda_2 \lambda_3 \lambda \cos \theta_i/2kT)$ and the corresponding distance jumped is $\lambda \cos \theta_i$, so that $\sum_i \lambda \cos \theta_i k_i \exp (f_s \lambda_2 \lambda_3 \lambda \cos \theta_i/2kT)$ is the velocity of the molecule due to all jumps into neighboring sites.

For Newtonian flow for which

$$f_s \lambda_2 \lambda_3 \lambda \cos \theta_i/2kT \ll 1 \tag{5.4}$$

expansion of the exponential term in Eq. (5.3) gives:

$$\eta_s = \frac{f_s}{\dot{s}} = f_s \bigg/ \left[\frac{\lambda}{\lambda_1} \sum_i k_i \left(\cos \theta_i + \frac{f_s \lambda_2 \lambda_3 \lambda \cos^2 \theta_i}{2kT} \right) \right] \tag{5.5}$$

Since the sites are randomly distributed over the solid angle, the first term in the summation of Eq. (5.5) has the value $\sum_i k_i \cos \theta_i = 0$ and the second term has the value:

$$\sum_i k_i \frac{f \lambda_2 \lambda_3 \lambda}{2kT} \cos^2 \theta_i = \kappa \frac{kT}{h} Z \frac{V - V_s}{V} (1 - e^{-\theta/T})$$

$$\times \exp - \left(\frac{a'E_s V_s}{(V - V_s)RT} \right) \times \frac{f \lambda_2 \lambda_3 \lambda}{6kT} \tag{5.6}$$

Using the relations, $\lambda_1 = \lambda_2 = \lambda_3 = \lambda$ and $\lambda^3 = \sqrt{2} V_s/N$, and putting Eq. (5.6) into Eq. (5.5), we obtain

$$\eta_s = \frac{Nh}{Z\kappa} \frac{V}{V_s} \frac{6}{\sqrt{2}} \frac{1}{(V - V_s)} \frac{1}{(1 - e^{-\theta/T})} \exp \left[\frac{a'E_s V_s}{(V - V_s)RT} \right] \tag{5.7}$$

The term η_g is derived from the kinetic theory of gases [7], i.e.,

$$\eta_g = \frac{2}{3d^2} \left(\frac{mkT}{\pi^3} \right)^{1/2} \tag{5.8}$$

where m is the molecular mass and d is the diameter of the molecules. Introducing Eqs. (5.7) and (5.8) into (5.2), we obtain the viscosity equation for the liquid:

$$\eta = \frac{Nh}{Z\kappa} \frac{6}{\sqrt{2}} \frac{1}{1 - e^{-\theta/T}} \frac{1}{V - V_s} \exp \left[\frac{a'E_s V_s}{(V - V_s)RT} \right] + \frac{V - V_s}{V} \frac{2}{3d^2} \left(\frac{mkT}{\pi^3} \right)^{1/2} \tag{5.9}$$

It is interesting to compare the viscosity formula in Eq. (5.9) with earlier procedures from rate theory. Formerly, the gas-like contribution to viscosity was neglected and the liquid was treated as the relaxation of a lattice structure. This is a reasonable approximation near the melting point but becomes less appropriate with rise in temperature. It is especially satisfactory that our viscosity equation for the liquid passes over naturally into the gas equation. The present equation also suggests that since $V - V_s$ becomes zero for the solid, the solid should become very viscous and further that nonequilibrium dislocations and vacancies must be chiefly responsible for the diffusion and plasticity shown by solids.

A great deal of experimental evidence shows that liquids subjected to stress relax by the same mechanism regardless of the nature of the stress. Thus, the activation energy for either ionic diffusion or conduction is ordinarily equal to the activation energy for viscous flow, indicating that the same elementary reactions are involved. Self-diffusion and viscous flow show the same activation energy and therefore involve the same elementary process.

We consider next the relation between viscosity and diffusion. The diffusion coefficient D is defined by the following equation:

$$J = -D \frac{dC}{dX} = CU \tag{5.10}$$

where J is the current density of matter and U and C are the velocity and concentration of matter, respectively. In Figure 5-2, we have drawn a typical

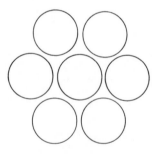

Figure 5-2 The hexagon normal to the direction of self-diffusion of the central molecules corresponds to a close packed liquid. The central molecules diffuses forward by shearing past nearest neighbors in random fashion.

cross section of a condensed phase normal to the direction of flow. We assume that the central molecule diffuses upward by randomly jumping past the six neighboring molecules. This is a necessary assumption, since the activation for the shear mechanism is involved in diffusion. The rate of shear is

$$\dot{s} = \frac{U}{\lambda_1} \tag{5.11}$$

where U is the forward velocity of the central molecule with respect to its neighbors. Further,

$$\eta = \frac{f}{\dot{s}} \tag{5.12}$$

where f, the shear stress, is the driving force on the central molecule divided by the sum, A, of the six nearest-neighbor shear areas; that is, $A = 6\lambda_2\lambda_3$ or in general $A = \xi\lambda_2\lambda_3$ where ξ is the number of nearest neighbors.

For diffusion, the force driving a molecule is

$$F = -\frac{d\mu}{dX} = -\frac{d}{dX}(kT \ln a + \mu_0)$$

$$= -kT \frac{d \ln a}{dX} = -kT \frac{d \ln a}{d \ln C} \frac{1}{C} \frac{dC}{dX} \tag{5.13}$$

Since the driving potential causing shear is the chemical potential

$$\mu = \mu_0 + kT \ln a \tag{5.14}$$

Here μ, μ_0, and a are respectively the chemical potential, the chemical potential at unit activity, and the activity. Also, we have

$$f = \frac{F}{A} = -\frac{kT}{\xi\lambda_2\lambda_3}\left(\frac{d \ln a}{d \ln C}\right)\frac{1}{C}\frac{dC}{dX} = \eta\dot{s} \tag{5.15}$$

Combining Eqs. (5.11), (5.12), and (5.15) with (5.10) yields

$$D = \frac{\lambda_1 kT}{\xi\lambda_2\lambda_3\eta}\frac{d \ln a}{d \ln C} \tag{5.16}$$

For self-diffusion $d \ln a / d \ln C \simeq 1$ and

$$D = \frac{kT}{\xi\dfrac{\lambda_2\lambda_3}{\lambda_1}\eta} \tag{5.17}$$

We expect values for ξ near six.

Li and Chang [8] have discussed the nature of ξ from a somewhat different point of view and also have compiled the available data on self-diffusion. An alternative procedure in deriving Eq. (5.16) is to follow Einstein in equating the Stokes hydrodynamic viscous drag on a sphere [9],

$$F = 6\pi r\eta U \frac{\beta r + 2\eta}{\beta r + 3\eta} \tag{5.18}$$

to the thermodynamic drag, Eq. (5.13). The resulting equation is solved for U, which when introduced into Eq. (5.10) leads to

$$D = \frac{kT}{6\pi r \eta} \frac{d \ln a}{d \ln C} \left(\frac{\beta r + 3\eta}{\beta r + 2\eta} \right) \qquad (5.19)$$

Except for β, the coefficient of sliding friction, all the notation has been defined previously. If there is no slip at the interface, i.e., if β is very large, the denominator in Eq. (5.19) is $6\pi r \eta$; if $\beta = 0$, the denominator becomes $4\pi r \eta$. It is very interesting to compare Eq. (5.19) with Eq. (5.16): $4\pi r$ is the circumference of the circle drawn through the centers of the surrounding molecules, while $\xi(\lambda_2 \lambda_3 / \lambda_1)$ is the sum of the length of straight lines which form the closed polygon joining the centers of the surrounding molecules. In conjunction with the viscosity equation in Eq. (5.9), Blomgren [10] obtained a slightly different form by applying the early viscosity equation from absolute reaction rate theory.

5.3 VISCOSITY AND DIFFUSION OF HARD SPHERE SYSTEMS

For the application of the viscosity and diffusion equations developed previously, we describe the rigid sphere system [11] and compare the calculated results with Enskog's theory. Equation (5.7) can be rewritten in an alternative manner:

$$\eta_s = \frac{V}{V_s} \frac{Nh}{Z\kappa} \frac{6}{\sqrt{2}} \frac{1}{V - V_s} f_{s_1} \exp (\varepsilon_0^{\ddagger}/kT) \quad \text{here} \quad \varepsilon_0^{\ddagger} = \frac{a' E_s V_s}{(V - V_s)RT} \qquad (5.20)$$

where f_{s_1} is the partition function of a solid in one degree of freedom, and $\varepsilon_0^{\ddagger}$ is the activation energy required for jumping. When Lennard-Jones and Devonshire's cell theory is considered instead of the Einstein oscillator partition function, f_{s_1} is written

$$f_{s_1} = \frac{(2\pi mkT)^{1/2}}{h} l_f \qquad (5.21)$$

where l_f is the free length. If in the above equation, the rigid sphere condition is introduced, i.e.,

$$E_s = \phi(a) = 0 \qquad (5.22)$$

$$l_f = 2(a - d) \qquad (5.23)$$

$$\kappa = 1 \qquad (5.24)$$

then, the viscosity for the rigid sphere becomes

$$\eta = (\pi m k T)^{1/2} N \frac{\left(\sqrt{2}\dfrac{V_s}{N}\right)^{1/3} - d}{V - V_s} + \frac{V - V_s}{V} \frac{2}{3d^2} \left(\frac{mkT}{\pi^3}\right)^{1/2} \quad (5.25)$$

In Figure 5-3 we compare the hard sphere viscosities calculated from Eq. (5.25) with those calculated from Enskog's theory. As shown in the figure, the two

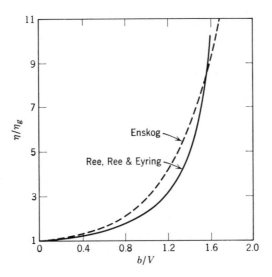

Figure 5-3 Comparison with Enskog's theory. (after Ree *et al* [11]).

theories agree quite well. The diffusion coefficient also may be calculated by means of Eq. (5.17) which was obtained by Eyring and Ree [5]:

$$D = \frac{kT}{\xi \dfrac{\lambda_2 \lambda_3}{\lambda_1} \eta} \quad (5.17)$$

where, for hexagonal packing, $\xi = 6$ and

$$\lambda_2 \lambda_3 = \frac{\sqrt{3}}{2} a^2, \quad \lambda_1 = \frac{\sqrt{3}}{2} a, \quad \lambda = a \quad (5.26)$$

$$\frac{a^3}{\sqrt{2}} = \frac{V_s}{N} \quad (5.27)$$

Here a is the distance between two nearest neighbors. Substituting Eqs. (5.26) and (5.27) into (5.16), we obtain

$$D = \frac{kT}{6\left(\dfrac{\sqrt{2}\,V_s}{N}\right)^{1/3} \eta}$$

(5.28)

In Table 5-1, we compare the diffusion coefficient calculated by using Eq. (5.28) with experimental data. Agreement between theory and experiment is quite good in general.

Table 5-1 Comparison of Calculated and Experimental Coefficients for Simple Liquids

Substance and V_s	T (°K)	$\eta \times 10^3$ (poise)	$D_{exp} \times 10^5$ (cm²/sec)	$D_{theo} \times 10^5$ (cm²/sec)
Ar	84.31 ± 0.13	2.82	2.07 ± 0.06	1.77
$V_s = 24.98$ (cm³/mole)	90	2.32	2.43	2.30
	100	1.7	3.54	3.48
	110	1.40	4.80	4.65
	120	1.14	6.06	6.23
	130	0.90	7.45	8.54
	140	0.69	8.72	12.0
	150	0.48	9.98	18.5
CH₄	100	1.45	3.0	3.7
$V_s = 31.06$ (cm³/mole)	110	1.18	4.2	5.0
	120	1.02	6.0	6.6
	130	0.94	7.8	8.5
C₆H₆	287.7	7.08	1.6	1.61
$V_s = 82.82$ (cm³/mole)	288.2	6.96	1.88 ± 0.01	1.64
	298.2	5.99	2.15 ± 0.05	1.98
	308.2	5.30	2.40 ± 0.03	2.31
	318.2	4.60	2.67 ± 0.06	2.75
CCl₄	289.2	8.88	1.41	1.31
$V_s = 87.1$ (cm³/mole)	308.2	7.70	1.75	1.56
	318.2	6.90	1.99	1.80

5.4 VISCOSITY IN DENSE GASES

In the early part of Chapter 4, the more general partition function of the solid-like degree of freedom in the dense gas region was discussed. Following Ree, Ree, and Eyring's treatment [12], f_{s_1} in Eq. (5.20) has the following form:

$$f_{s_1} = \frac{1 - e^{-l\theta/T}}{1 - e^{-\theta/T}} + e^{-l\theta/T}\frac{(2\pi mkT)^{1/2}}{h}\left[\left(\frac{V_s}{N}\right)^{1/3} - \left(\frac{b}{4N}\right)^{1/3}\right]$$

(5.29)

This expression for f_{s_1} assumes that when the vibrational energy of a molecule in the liquid reaches one-third of the heat of vaporization, $l\theta = E_s/3R$, the degree of freedom becomes gas-like with a free length $[(V_s/N)^{1/3} - (b/4N)^{1/3}]$ where b is van der Waals constant, which when divided by 4 indicates the hard volume of the molecule. In this particular system under higher pressure, the pressure effects also have been introduced for the expression of V_s and of η_s. Introducing Eq. (5.29) into Eq. (5.20), and considering the effect of pressure, we obtain:

$$\eta = \frac{Nh}{Z\kappa} \frac{6}{\sqrt{2}} \frac{1}{V - V_s} \left[\frac{1 - e^{-E_s/3RT}}{1 - e^{-\theta/T}} + e^{-E_s/3RT} \frac{(2\pi mkT)^{1/2}}{h} \right.$$

$$\left. \times \left[\left(\frac{V_s}{N}\right)^{1/3} - \left(\frac{b}{4N}\right)^{1/3} \right] \right]$$

$$\times \exp\left(\frac{a' E_s V_s}{(V - V_s)RT} \right) \cdot \exp \frac{P(V - V_s)}{RT} + \frac{V - V_s}{V} \frac{2}{3d^2} \left(\frac{mkT}{\pi^3}\right)^{1/2} \quad (5.30)$$

Here, the correction of the V_s of the solid-like structure at pressure P is given by the equation

$$V_{sp} = V_s(1 - \beta \Delta P)$$

where β is the coefficient of compressibility for the solid and V_s is the solid volume at the melting point. The last exponential term in Eq. (5.30) for η_s has been introduced to take care of the pressure effect. Letting $\kappa = 0.375$,

Table 5-2   Viscosities (milipoise) of Argon Under its Vapor Pressure

T (°K)	η_{calc}	η_{obs}	Δ (%)
84.25	2.91	2.82	+3.2
86.90	2.60	2.56	+1.6
90	2.32	2.32	+0.0
111	1.35	1.37	−2.2
133.5	0.81	0.77	+2.6
143	0.65	0.63	+3.2
149	0.55	0.50	+2.0

$a = 0.00534$, and $Z = 12$, the viscosity of argon under vapor pressure at various temperatures were calculated and are shown in Table 5-2. The agreement with experiment is satisfactory over the whole range.

Zhandanova [13] measured the viscosities of argon at constant volume in both the liquid and gaseous ranges. The experimental data shown in Figure 5-4 are due to her, where the fluidity $\phi = \eta^{-1}$ is plotted against temperature.

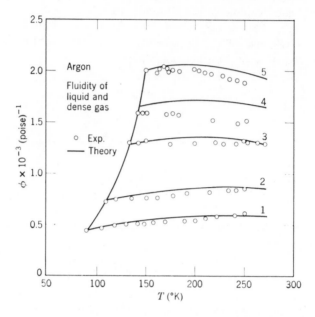

Figure 5-4   Fluidities (poise^{-1}) of argon at constant volume versus temperatures (°K). Curve 1 at $V = 29.66$ cc; curve 2 at $V = 33.416$ cc; curve 3 at $V = 40.59$ cc; curve 4 at $V = 46.48$ cc; curve 5 at $V = 52.78$ cc. (after Ree *et al.* [12])

Curves 1 to 5 are the values calculated from Eq. (5.30). The agreement with experiment is quite good over the whole range.

5.5   VISCOSITY IN LIQUIDS AND LIQUID MIXTURES

As a further check of the applicability of the viscosity equation, Eq. (5.9), the viscosities of the following liquids have been calculated so far: liquid argon [14]; molten metals [15]; p- and m-xylenes [16]; liquid fluorine [17]; liquid oxygen [18]; fused salt [19, 20]. Some of the calculated results are shown in Figure 5-5 and Table 5-3 and are compared with experiment. The results are quite satisfactory.

If the model discussed for viscosity is reasonable, it can be applied to mixtures. The binary liquid partition functions discussed in Chapter 4 can be

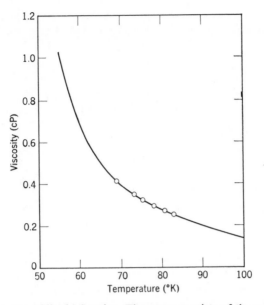

Figure 5-5 Viscosity of liquid fluorine. The curve consists of the calculated values and the circles indicate the experimental points. (after Thomson *et al.* [17])

Table 5-3 Viscosity of Molten Salts

Substances	$T\,(^\circ K)$	η_{calc} (cp)	η_{obs} (cp)
NaCl	1074	1.1601	—
($d = 2.36$ Å)	1080	1.1422	—
	1200	0.8877	0.9000
	1290	0.7749	—
	1500	0.6232	—
	1738	0.5257	—
KCl	1043	1.2202	—
($d = 2.67$ Å)	1050	1.1932	—
	1100	1.0328	1.000
	1200	0.8248	0.800
	1500	0.5499	—
	1680	0.4680	—
KBr	1007	1.3223	—
($d = 2.82$ Å)	1020	1.2748	1.180
	1100	1.0534	0.950
	1150	0.9576	0.870
	1200	0.8824	—
	1653	0.5637	—

used to develop the viscosity equations for mixtures. At the present time, work is beginning on this subject. The viscosity formula being considered is

$$\eta = \frac{hN}{r}\frac{6}{\sqrt{2}}\frac{1}{(1 - e^{-\theta_1/T})^{X_1}}\frac{1}{(1 - e^{-\theta_2/T})^{X_2}}\frac{1}{V - V_s}\exp\frac{bE_sV_s}{RT(V - V_s)}$$

$$+ \frac{V - V_s}{V}\left[\frac{2}{3d_1^2}\left(\frac{m_1kT}{\pi^3}\right)^{1/2}X_1 + \frac{2}{3d_2^2}\left(\frac{m_2kT}{\pi^3}\right)^{1/2}X_2\right] \qquad (5.31)$$

Here $r = Z\kappa$, $m_1 = M_1/N$, and $m_2 = M_2/N$

$$E_s = X_1^2 E_{s_1} + X_2^2 E_{s_2} + 2X_1X_2(E_{s_1}E_{s_2})^{1/2}(1 + \delta E_s),$$

$$V_s = V_{s_1}X_1 + V_{s_2}X_2,$$

$$b = b_1X_1 + b_2X_2,$$

$$r = r_1X_1 + r_2X_2$$

Suffixes 1 and 2 indicate the quantities of the species 1 and 2 in the mixture, and X denotes the mole fraction of the species. For the derivation of Eq. (5.31), the interaction between the viscosities of the gas-like degree of freedom of species 1 and 2 is neglected. It is further assumed that the parametric values r, V_s, E_s, and b are only concentration-dependent, and no extra adjustable parameters are used in Eq. (5.31). The results so far obtained are encouraging.

5.6   THEORY OF THERMAL CONDUCTIVITY

Just as in the case of the viscosity, we represent the thermal conductivity of liquids [21] as follows:

$$\kappa = \left(\frac{V_s}{V}\right)\kappa_s + \left(\frac{V - V_s}{V}\right)\kappa_g \qquad (5.32)$$

where κ is the liquid thermal conductivity, κ_s is that of the solid-like degrees of freedom, and κ_g is that of gas-like degrees of freedom. For the thermal conductivity of gases, κ_g, the general results [22] are usually expressed as

$$\kappa_g = \varepsilon\eta_g C_{vg} \qquad (5.33)$$

where ε is a pure number, η_g the viscosity, and C_{vg} the specific heat at constant volume. Since the contribution of gas thermal conductivity to the liquid thermal conductivity is generally small, it will suffice to use the ideal gas equations. Thus

$$\kappa_g = (9\gamma - 5)/4 \cdot \eta_g C_{vg} \qquad (5.34)$$

where γ is the ratio of specific heat at constant pressure to that at constant volume, and η_g and C_{vg} are the viscosity and the heat capacity at constant volume of an ideal gas, respectively. For a more rigorous treatment of κ_g, one should refer to the literature [23, 24]. For the calculation of κ_s, Lin, Eyring, and Davis [21] used the phonon theory first developed by Peierls [25, 26] and then modified by Klemens [27] and Callaway [28]. The expression for κ_s is

$$\kappa_s = \frac{C_s^2 R}{V_s(A\omega_e^4 + BT\omega_e^2)} \times \frac{(\hbar\omega_e/kT)^2 e^{\hbar\omega_e/kT}}{(e^{\hbar\omega_e/kT} - 1)^2} \tag{5.35}$$

where ω_e is the Einstein characteristic frequency; $A\omega_e^4$ represents the scattering by lattice-imperfection; the term $BT\omega_e^2$ includes the normal and the Umklapp processes; and C_s is the sound velocity of solids. For the temperature of interest, we obtain the following relationship:

$$\frac{\hbar\omega_e}{kT} < 1 \tag{5.36}$$

Hence

$$\frac{(\hbar\omega_e/kT)^2 e^{\hbar\omega_e/kT}}{(e^{\hbar\omega_e/kT} - 1)^2} \simeq 1 \tag{5.37}$$

Substituting Eqs. (5.34) and (5.35) with (5.37) into Eq. (5.32), we obtain

$$\kappa = \frac{RC_s^2}{V(A\omega_e^4 + BT\omega_e^2)} + \frac{V - V_s}{V}\frac{9\gamma - 5}{4}\eta_g C_{vg} \tag{5.38}$$

The first term on the right-hand side of Eq. (5.38) represents the contribution to thermal conduction due to the vibration of molecules near their equilibrium position, and the second term, the contribution due to the random motion of molecules. In Figures 5-6 and 5-7, the values of κ calculated from

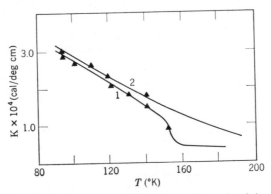

Figure 5-6 Temperature dependence of the thermal conductivity of argon: full curves, experimental results; points, calculated values; curve 1, $P = 48$ atm, curve 2, $P = 120$ atm. (after Lin *et al.* [21])

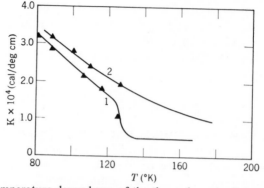

Figure 5-7 Temperature dependence of the thermal conductivity of nitrogen: full curves, experimental results; points, calculated values; curve 1, $P = 33.5$ atm; curve 2, $P = 134$ atm. (after Lin *et al.* [21])

Eq. (5.38) at constant pressure are plotted against T °K, where $A = 2.50 \times 10^{-40}$ and $B = 5.52 \times 10^{-16}$ for argon and $A = 1.82 \times 10^{-40}$ and $B = 5.64 \times 10^{-16}$ for nitrogen. These values are found empirically by using experimental thermal conductivity at two given temperatures. The resulting thermal conductivity for solid-like degrees of freedom also gives good results for the solid below the melting point. Agreement between theory and experiment is quite good.

To consider the effects of pressure on the liquid thermal conductivity, we need to estimate the dependence of C_s on the pressure,

$$C_s = (\gamma_s/\rho_s \beta_T)^{1/2} \tag{5.39}$$

where γ_s is the ratio of specific heats of solids, ρ_s is the density of solids, and β_T the isothermal compressibility coefficient of solids. In Figures 5-8 and 5-9 the liquid thermal conductivity versus pressure below 140 atm is shown. For small ranges of pressure, κ is linear with respect to pressure.

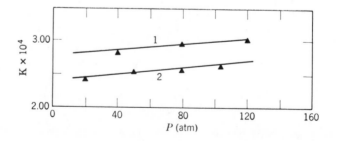

Figure 5-8 Pressure effect on κ_1 of argon: ———, experimental results; ▲, calculated values; curve 1, $T = 93.3$°K; curve 2, $T = 106.1$°K. (after Lin *et al.* [21])

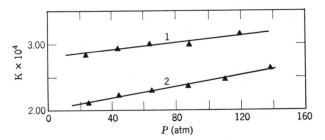

Figure 5-9   Pressure effect on κ_1 of nitrogen: ————, experimental results; ▲, calculated values: curve 1, $T = 87.7°K$; curve 2, $T = 105.6°K$. (after Lin *et al.* [21])

REFERENCES

[1]   J. G. Kirkwood, *J. Chem. Phys.*, **14**, 180 (1946).
[2]   (a) S. Chapman and T. G. Cowling, *The Mathematical Theory of Non-uniform Gases*, Cambridge University Press, Cambridge, 1939, Chapter 16; (b) *ibid.*, Chapter 10.
[3]   H. Eyring, D. Henderson, B. Stover, and E. Eyring, *Statistical Mechanics and Dynamics*, Wiley, New York, 1964.
[4]   R. Marchi and H. Eyring, *J. Chem. Educ.*, **40**, 526 (1963).
[5]   H. Eyring and T. Ree, *Proc. Natl. Acad. Sci. (U.S.)* **47**, 526 (1961).
[6]   H. Eyring, *J. Chem. Phys.*, **4**, 283 (1936).
[7]   W. J. Moore, *Physical Chemistry*, Prentice-Hall, Englewood Cliffs, N.J., 1959, p. 178.
[8]   J. C. M. Li and P. Chang, *J. Chem. Phys.*, **23**, 518 (1955).
[9]   H. Lamb, *Hydrodynamics*, Dover, New York, 1932, 6th ed., p. 602, eq. 38.
[10] G. E. Blomgren, *Ann. New York Acad. Sci.*, 781 (1960).
[11] T. S. Ree, T. Ree, and H. Eyring, *J. Phys. Chem.*, **68**, 3262 (1964).
[12] T. S. Ree, T. Ree, and H. Eyring, *Proc. Natl. Acad. Sci. (U.S.)*, **48**, 501 (1962).
[13] N. F. Zhandanova, *Soviet Physics, JETP*, **4**, 749 (1957).
[14] E. J. Fuller, "Significant Liquid Structure," Ph.D. Thesis, University of Utah, 1960.
[15] C. M. Carlson, H. Eyring, and T. Ree, *Proc. Natl. Acad. Sci. (U.S.)*, **46**, 649 (1960).
[16] M. S. John, J. Grosh, T. Ree, and H. Eyring, *Proc. Natl. Acad. Sci. (U.S.)*, **54**, 1419 (1965).
[17] T. R. Thomson, H. Eyring, and T. Ree, *J. Phys. Chem.*, **67**, 2701 (1963).
[18] K. C. Kim, W. C. Lu, T. Ree, and H. Eyring, *Proc. Natl. Acad. Sci. (U.S.)*, **57**, 861 (1967).
[19] W. C. Lu, T. Ree, V. G. Gerrard, and H. Eyring, *J. Chem. Phys.*, in press.
[20] M. S. Jhon, G. Clemena, and E. R. Van Artsdalen, *J. Phys. Chem.*, **72**, 4155 (1968).
[21] S. H. Lin, H. Eyring, and W. J. Davis, *J. Phys. Chem.*, **68**, 3021 (1964).
[22] L. B. Loeb, *Kinetic Theory of Gases*, McGraw-Hill, New York, 1927, Chapter 6.
[23] J. O. Hirschfelder, C. F. Curtiss, and R. B. Bird, *Molecular Theory of Gases and Liquid*, Wiley, New York, 1954.
[24] S. Chapman and T. G. Cowling, *The Mathematical Theory of Non-uniform Gases*, Cambridge University Press, London, 1953.
[25] R. Peierls, *Ann. der Phys. Lpz.*, **3**, 1055 (1929).
[26] R. Peierls, *Quantum Theory of Solids*, Oxford University Press, 1950.
[27] P. G. Klemens, *Proc. Roy. Soc.*, **A208**, 108 (1951).
[28] J. Callaway, *Phys. Rev.*, **113**, 1046 (1959).

chapter 6

THE SURFACE TENSION
OF LIQUIDS

6.1 INTRODUCTION

Several theories have been proposed for the surface tension of liquids, ranging from fairly simple treatments in terms of easily visualized physical concepts, such as the theory of significant structures in liquids [1], to more detailed dynamic calculations. Theories in the latter category have been thoroughly reviewed by Onno and Kondo [2].

According to the significant structure theory of liquids, there are two typical approaches for developing a theory of surface tension. One is an iteration procedure which calculates the sum of the contributions of successive molecular layers; the other is a monolayer approximation which yields a simplified approximate calculation.

6.2 THE ITERATION METHOD

Since the partition function which has been used reduces to the gas partition function for $V \gg V_s$, it is to be expected that the partition function will be applicable to the surface layer as well as to the bulk liquid. Consequently, the fundamental aspects of the expression for the partition function are retained: As before, the significant structures are (a) the solid-like degrees of freedom, (b) positional degeneracy arising from the presence of the holes, and (c) gas-like degrees of freedom. The principal change required in the partition function, Eq. (3.9), for monatomic liquids arises from the differences in the forces acting on a surface molecule and a bulk molecule:

$$f_N = \left\{ \frac{e^{E_s/RT}}{(1 - e^{-\theta/T})^3} (1 + n_h e^{-aE_s/n_h RT}) \right\}^{NV_s/V} \left\{ \frac{(2\pi mkT)^{3/2}}{h^3} \frac{eV}{N} \right\}^{N\left(\frac{V-V_s}{V}\right)} \quad (3.9)'$$

This difference is related to the value of E_s in Eq. (3.9)'. The correction for E_s is obtained by an iteration method. For a simple close-packed liquid such as argon it is easy to visualize that a given molecule has six nearest neighbors in

96

the same layer, three nearest neighbors in the layer below, and three in the layer above, giving a total of twelve nearest neighbors. Thus, in the surface of a liquid, the value of E_s should be different from that of the bulk liquid. The energy of a molecule in the ith surface molecular layer is then given by the equation:

$$E_{s_i} = E_s \left(\frac{6}{12} \frac{\rho_i}{\rho_i} + \frac{3}{12} \frac{\rho_{i+1}}{\rho_i} + \frac{3}{12} \frac{\rho_{i-1}}{\rho_i} \right) \tag{6.1}$$

where E_{s_i} is the corrected value of E_s for the ith layer, and ρ_i, ρ_{i+1}, and ρ_{i-1} are the densities of the ith layer and the layers immediately below and above. The thickness of a molecular layer can be deduced from the cubic closest packing structure. The face of the lattice which corresponds to the lowest surface free energy is the one exposed. Since cubic closest packing can be referred to as a face centered cubic lattice, the distance d between two successive planes of (III) which is the exposed face [3, 4] is given by

$$d = \frac{a}{\sqrt{3}} \tag{6.2}$$

where a is the lattice parameter. Since four molecules are involved in a unit cell of the lattice, the volume a^3 is equal to $4(V_s/N)$. Therefore Eq. (6.2) becomes

$$d = \left(\frac{4V_s}{N} \right)^{1/3} \bigg/ \sqrt{3} = 0.9165 \left(\frac{V_s}{N} \right)^{1/3} \tag{6.3}$$

The surface tension of a liquid γ is the excess Gibbs free energy per unit area over that of the bulk liquid. Therefore, the surface tension is given by

$$\gamma = \sum_i (G_{s_i} - G_l) \times \frac{d}{V_i} = \sum_i (G_{s_i} - G_l) \times 0.9165 \left(\frac{V_s}{N} \right)^{1/3} \bigg/ V_i \tag{6.4}$$

where G_{s_i} is the molar Gibbs free energy of the ith surface layer; G_l the molar Gibbs free energy of the bulk liquid; V_i the molar volume of the ith layer; and $0.9165(V_s/N)^{1/3}/V_i$ the number of moles per unit area of the ith layer. The molar free energy of the ith surface layer is calculated from the partition function Eq. (3.9) by substituting the value of E_{s_i} of Eq. (6.1). The corresponding molar volume V_i is determined by plotting A_i versus x ($= V/V_s$), and by drawing a tangent whose slope is the vapor pressure; then V_i is the volume at the tangential point. (See Chapter 4.)

To calculate the value of E_{s_i} of Eq. (6.1), we have to know the densities of the $(i - 1)$th, the ith, and the $(i + 1)$th layers. Unfortunately, we only know the densities of the bulk liquid and gas phases. Thus, an iteration technique is applied which will be explained below. For calculating the first layer, the preliminary assumption used is that the density of the layer immediately below, ρ_2, is equal to the bulk liquid density, ρ_l, and that the density of the first gaseous layer, ρ_0, is equal to the gas density, ρ_g. Introducing these approximations into Eq. (6.1) gives

$$E_{s_1} = E_s \left[\frac{1}{2} + \frac{1}{4}\frac{\rho_l}{\rho_1} + \frac{1}{4}\frac{\rho_g}{\rho_1} \right] \tag{6.5}$$

where

$$\frac{\rho_g}{\rho_1} = \exp \frac{E_s}{2RT} \left(\frac{T}{T_c} - 1 \right) \tag{6.5a}$$

The relationship given by Eq. (6.5a) can be considered as follows. A molecule in the top liquid layer possesses six nearest neighbors in the same plane, three in the layer immediately below, but none in the gas layer, while a molecule in the first gas layer has only the three neighbors in the top liquid layer. Thus, the difference in energy of a surface and of a first gaseous layer molecule is the difference between nine bonds and three bonds, or

$$\frac{9-3}{12} E_s = \frac{E_s}{2}$$

We can thus write Eq. (6.5a) for the ratio of densities of the last gas layer and the first liquid layer. This equation satisfies the requirement that ρ_g/ρ_1 approach unity as T goes to T_c. For the first step of the calculation, we assume that the density of the top surface layer, ρ_1, and that of the layer immediately below, ρ_2, are equal to the bulk liquid density, ρ_l. Introducing these approximations into Eq. (6.5) gives

$$E'_{s_1} = E_s \left(\frac{3}{4} + \frac{1}{4}\exp \frac{E_s}{2RT} \left(\frac{T}{T_c} - 1 \right) \right) \tag{6.6}$$

as the first estimate of E_s for the surface layer. This value is used to evaluate the Helmholtz free energy A versus x and leads to a new ρ_1 for the surface layer. The E''_{s_1} is calculated by using ρ_1 in Eq. (6.1) and the relation in (6.5a) and the procedure is repeated until ρ_1 converges to a constant value. For the second layer calculation, the approximation for E'_{s_2}, which is used for the first calculation, is

$$E'_{s_2} = E_s \left(\frac{3}{4} + \frac{1}{4}\frac{\rho_1}{\rho_l} \right) \tag{6.7}$$

where E'_{s_2} is the first corrected value of E_s and ρ_1 is the density of the first layer, which was calculated previously.

By a similar procedure to that described in the first iteration, the value of ρ_2 is found, and E''_{s_2} is calculated from Eq. (6.1). A constant value of ρ_2 is obtained again by iteration. The value of ρ_2 is then used to recalculate E_{s_1} and the recalculated value of ρ_1 is used to recalculate ρ_2. This iteration is repeated until one obtains constant values of $\rho_1, \rho_2, \ldots, \rho_n$ to the nth layer. In the calculation of the surface tension of the liquid, the contribution of the layer below the third is small. In the case of argon, the contribution of the third layer is negligible. The parametric values used and the calculated results of the surface tension of argon obtained from the above scheme are listed and compared with the observed values [5] in Tables 6-1 and 6-2. The results are

Table 6-1 Parameters for Liquid Argon

n	a	E_s (cal/mole)	V_s (cm³/mole)	θ (°K)
10.8	0.00534	1888.6	24.98	60.0

Table 6-2 Surface Tension of Argon

T (°K)	V (cm³/mole)	% Contribution of each layer			γ_{calc} (dyne/cm)	γ_{obs} (dyne/cm)	Δ error
		1st	2nd	3rd			
83.85 (T_m)	28.90	87.16	12.84	0	13.78	13.5	2.07
85.5	29.08	86.45	13.55	0	13.43	13.1	2.52
87.29 (T_b)	29.33	85.87	14.13	0	13.02	12.6	3.33
90.0	29.80	85.35	14.65	0	12.22	11.9	2.69

quite satisfactory. The Eötvös constant, K_E, which is defined by $K_E = \gamma V^{2/3} / (T_c - T)$, is calculated using the calculated values of γ and the observed volume, V, given in Table 6-2. The averaged theoretical K_E's are compared with the observed ones in Table 6-3. The agreement is very good. In Figure

Table 6-3 The Eötvös Constant for Argon

T (°K)	K_E (calc.)	K_E (calc.) Average	K_E (obs.)
83.85 (m.p)	1.903	1.916	2.02
85.5	1.916		
87.29 (b.p)	1.926		
90.0	1.911		

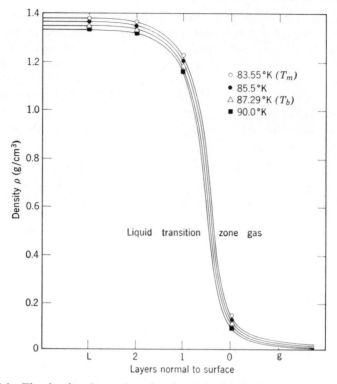

Figure 6-1 The density change in going from the liquid phase through the transition layers into the gas phase for argon at four temperatures.

6-1 the calculated density change from the liquid phase to the gas phase is plotted at the four temperatures. It can be seen that the range of densities decreases as the temperature increases; this is natural since the density curve approaches a horizontal line as T approaches T_c.

6.3 SURFACE TENSION OF NONPOLAR POLYATOMIC LIQUIDS

If the partition functions of the liquid have been determined, the evaluation of the surface tension can be obtained by using the iteration technique discussed previously. In this manner, the surface tensions of the following liquids have been tested successfully: inorganic liquids including

halogens [6], halogen halides [7], oxygen [8], nitrogen [6a], hydrogen [9], boron hydrides [10], molten mercuric halides [11]; and

organic liquids comprising

methane [6a], ethane [12], benzene [6a], carbon tetrachloride [13].

Exemplifying the model the results of nitrogen and methane are given in Tables 6-4 and 6-5. There is good agreement between experiment and theory.

Table 6-4 Parameters for Liquid Nitrogen and Methane

	n	a	E_s (cal/mole)	V_s (cm³/mole)	θ (°K)
N₂	10.94	0.05017	1538.0	29.15	53.74
CH₄	11.05	0.04024	2201.0	31.06	71.34

Table 6-5 Surface Tension of Liquid Nitrogen and Methane

T (°K)	% Contribution of each layer			γ_{calc} (dyne/cm)	γ_{obs} (dyne/cm)	Δ %
	1st	2nd	3rd			
			Nitrogen			
63.30 (m.p)	90.24	9.76	0.00	11.78	12.05	−2.24
68.41	89.18	10.82	0.00	10.63	10.89	−2.39
77.34 (b.p)	85.22	13.52	1.26	8.73	8.91	−2.02
			Methane			
90.65 (m.p)	91.25	8.75	0.00	17.14	18.20	−5.82
99.67	90.20	9.80	0.00	15.30	16.24	−5.79
111.67 (b.p)	87.09	11.91	1.00	13.01	13.70	−5.04
123.15	84.34	14.26	1.40	10.73	11.34	−5.38

The partition functions [14] for liquid nitrogen and methane are given by

$$f_{N_2} = \left\{ \frac{e^{E_s/RT}}{(1 - e^{-\theta/T})^3} \frac{8\pi^2 IkT}{2h^2} \frac{1}{1 - e^{-h\nu/kT}} \left[1 + n_h e^{-aE_s/n_h RT}\right] \right\}^{NV_s/V}$$
$$\times \left\{ \frac{(2\pi mkT)^{3/2}}{h^3} \frac{eV}{N} \frac{8\pi^2 IkT}{2h^2} \frac{1}{1 - e^{-h\nu/kT}} \right\}^{N\left(\frac{V-V_s}{V_s}\right)} \quad (6.8a)$$

and

$$f_{CH_4} = \left\{ \frac{e^{E_s/RT}}{(1 - e^{-\theta/T})^3} \frac{8\pi^2 (8\pi^3 ABC)^{1/2}(kT)^{3/2}}{12h^3} \right.$$
$$\left. \times \prod_{i=1}^{9} \frac{1}{1 - e^{-h\nu_i/kT}} \left[1 + n_h e^{-aE_s/n_h RT}\right] \right\}^{N\frac{V_s}{V}}$$
$$\times \left\{ \frac{(2\pi mkT)^{3/2}}{h^3} \frac{eV}{N} \frac{8\pi^2 (8\pi^3 ABC)^{1/2}(kT)^{3/2}}{12h^3} \prod_{i=1}^{9} \frac{1}{1 - e^{-h\nu_i/kT}} \right\}^{N\left(\frac{V-V_s}{V_s}\right)}$$
$$(6.8b)$$

where the notation was defined in Chapter 4. Recently, a generalized scheme for improving the iteration method was developed by Jhon et al. [15]. According to their scheme, the partition function for the surface layers of a nonpolar liquid are corrected not only for the ground state energy of the solid-like molecules (E_s), as was done by Chang et al. [1a], but also for the number of nearest neighbors around molecules (n) and the Einstein characteristic temperature (θ), on the basis that the partition function for the surface layer has the same form as for the bulk liquid. The iteration technique is similar to that of Chang et al. [1a].

6.4   SURFACE TENSION OF POLAR LIQUIDS

The above method of obtaining the surface tension has been successfully applied to nonpolar liquids, but the deviations from the experimental values using this method are large in the case of polar liquids such as water, alcohol, and liquid ammonia. But the model, which takes account of orientation at the surface, leads to results in agreement with experiment. According to this model, since the molecules in the top surface layer are in an asymmetric field, due to the concentration gradient between liquid and gas as well as to the dipole moment of surrounding molecules, these surface molecules will orient in a way to reduce their free energy and only to a lesser extent in the opposite direction. The main contribution to the abnormal surface tension for polar liquids comes from this orientational effect, but the changes in density within a few molecular diameters of the interface also makes some contribution. The above considerations were introduced into the partition function of the top layer, in which the additional orientation in the partition function is expressed as $(kT/\mu X) \sinh (\mu X/kT)$ [16], where μ is the dipole moments of the molecule and X is the field strength. The partition functions below the top layer are assumed to equal that of the bulk liquid.

Table 6-6   Surface Tension of Light Water

Temperature ($^\circ$K)	γ_{calc} (dyne/cm)	γ_{obs} (dyne/cm)	Δ %
273.15	75.96	75.6	-0.71
278.15	74.00	74.9	-1.20
288.15	72.97	73.49	-0.71
293.15	72.49	72.75	-0.36
303.15	71.12	71.18	-0.08
313.15	69.67	69.56	-0.16
353.15	62.92	62.6	0.51
373.15	59.09	58.9	0.32

This idea has been tested for the following polar liquids with success: liquid water and heavy water [17], liquid methanol [12], liquid hydrazine [10a], liquid methyl halides [12], and liquid ammonia [18]. The calculated results for liquid water [17b] are listed in Table 6-6. The surface tension and its temperature dependence give excellent agreement with experiment. In this calculation, no shift of the equilibrium constant of the two structures of ice-like molecules in liquid water was assumed. (See Chapter 4.) The partition function of only the first layer f_1 is different from that of the bulk liquid and is given by the following equation:

$$f_1 = \left\{ \frac{\exp(E_{s_1}/RT)}{(1 - \exp(-\theta_1/T))^5} f_R' \prod_{i=1}^{3} \frac{1}{1 - \exp(-hv_i/kT)} \right.$$

$$\times (K)^{\frac{K}{(1+K)q}} \left[1 + \exp\left(-\frac{aE_s V_s}{(V - V_s)RT} \right) \right] \bigg\}^{(V_s/V_1)N}$$

$$\times \left\{ \frac{(2\pi mkT)^{3/2}}{h^3} \frac{eV_1}{N} \frac{8\pi^2(8\pi^3 ABC)^{1/2}(kT)^{3/2}}{2h^3} \frac{kT}{\mu X} \right.$$

$$\times \sinh \frac{\mu X}{kT} \prod_{i=1}^{3} \frac{1}{1 - \exp(-hv_i/kT)} \bigg\}^{\frac{(V_1 - V_s)N}{V_1}} \quad (6.9)$$

Here

$$f_R' = \frac{2\pi(2\pi IkT)^{1/2}}{2h} \frac{kT}{\mu X} \sinh \frac{\mu X}{kT}$$

The suffix appearing in f_1 indicates the first layer and the other notation is similar to that of the bulk liquid. In this calculation, $\mu X = 14.16 \times 10^{-14}$ erg for H_2O was used.

6.5 THE MONOLAYER APPROXIMATION

The iteration procedure for calculating surface tension of a liquid is very useful in its application to various liquids ranging from inert gases to polar liquids. However, this approach fails to provide a simple equation for calculating the surface tension since, after all, an iteration procedure is needed in any satisfactory calculation of the contribution to the Gibbs free energy of successive layers. However, to derive an equation for the surface tension in closed form, Ree, Ree, and Eyring [1b] made the approximation that the dividing surface between a liquid and its vapor phase is a monomolecular layer, in which a molecule has a free volume larger than for an interior molecule and a potential energy less than for the latter. There is some evidence [19] that the boundary between a liquid and its vapor is sharp and that the boundary extends over only about one molecular layer. This is a fairly reasonable approximation for simple liquids, but is an oversimplification for an H-bonded associated liquid such as water [17]. By introducing the approximation into the significant structure theory of the liquid, the partition function in Eq. (6.10) is derived. For the construction of the partition function, Ree *et al.* [1b] used the Lennard-Jones and Devonshire cell model instead of the Einstein oscillator model for the solid-like degree of freedom. Then,

$$
\begin{aligned}
f = & \left\{ \frac{(2\pi m k T)^{3/2}}{h^3} v'_f \left(\exp - \frac{Z\psi'(o)}{2kT} \right) \right. \\
& \left. \times \left[1 + \frac{n(V - V_s)}{V_s} \exp \frac{a'\psi'(o)V_s}{(V - V_s)kT} \right] J(T) \right\}^{N'V_s/V} \\
& \times \left\{ \frac{(2\pi m k T)^{3/2}}{h^3} v_f \left(\exp - \frac{Z\psi(o)}{2kT} \right) \right. \\
& \left. \times \left[1 + \frac{n(V - V_s)}{V_s} \exp \frac{a'\psi(o)V_s}{(V - V_s)kT} \right] J(T) \right\}^{N''V_s/V} \\
& \times \left\{ \frac{(2\pi m k T)^{3/2}}{h^3} \frac{eV}{N} J(T) \right\}^{(\frac{V - V_s}{V})N}
\end{aligned}
\tag{6.10}
$$

Here, the assumption has been made that the solid-like and gas-like molecules are randomly distributed in the surface as well as in the bulk liquid. In Eq. (6.10), $J(T)$ is the partition function for the internal degrees of freedom of a molecule; m is the mass of the molecule; h is the Planck constant; k is the Boltzmann constant; and v_f, the free volume of a molecule; $\psi(o)$ is the potential energy possessed by a molecule as it vibrates about its equilibrium position; Z is the coordination number of the molecule; the single prime (except for the parameter a) indicates surface quantities; and $N' + N'' = N$ where N' and N'' are the number of molecules in the surface and in the bulk liquid, respectively.

The surface tension, γ, is calculated from the Helmholtz free energy, A, as follows:

$$
\gamma = \left(\frac{\partial A}{\partial \Omega} \right)_{N, V, T} = \omega^{-1} \left(\frac{\partial A}{\partial N_c} \right)_{N, V, T}
\tag{6.11}
$$

where Ω is the surface area and N_c, the total number of sites available on the surface, is related to N by the relation $N' = (V_s/V)N_c$, since a random distribution of holes is assumed. The symbol ω represents the area occupied by one molecule and is equal to $(\sqrt{3}/2)a^2$ for close packing, where a is the nearest neighbor distance, and is given by $(\sqrt{2} V_s/N)^{1/3}$. Hence, by substituting

Eq. (6.10) into the relation, $A = -kT \ln f$, and by using Eq. (6.11), the surface tension is

$$\gamma = \frac{2}{\sqrt{3}} \left(\frac{N}{\sqrt{2} \, V_s} \right)^{2/3} \left(\frac{V_s}{V} \right)^2 \left[\frac{Z}{2} \psi'(o) - \frac{Z}{2} \psi(o) - kT \ln \frac{v_f'}{v_f} \right] \qquad (6.12)$$

In the derivation of Eq. (6.12), it was assumed that

$$\frac{\left\{ 1 + n\left(\dfrac{V - V_s}{V_s} \right) \exp\left(\dfrac{a'\psi'(o)V_s}{(V - V_s)kT} \right) \right\}}{\left\{ 1 + n\left(\dfrac{V - V_s}{V_s} \right) \exp\left(\dfrac{a'\psi(o)V_s}{(V - V_s)kT} \right) \right\}} \simeq 1$$

In order to obtain the value of v_f'/v_f and $\psi'(o) - \psi(o)$, Ree *et al.* [1b] assumed a square well potential and a Lennard-Jones potential. (See Figure 6-2.) The width of the square well for the motion toward the bulk liquid is

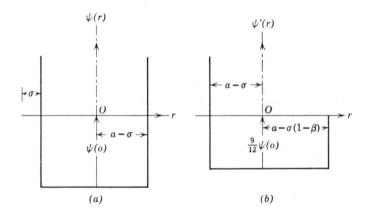

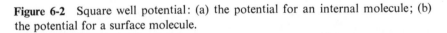

Figure 6-2 Square well potential: (a) the potential for an internal molecule; (b) the potential for a surface molecule.

$(a - \sigma)$, whereas that for the motion toward the vapor phase is $a - \sigma(1 - \beta)$. Hence v_f and v_f' are written as

$$v_f = \int_0^{a-\sigma} 4\pi r^2 \, dr = \frac{4}{3} \pi (a - \sigma)^3 \qquad (6.13a)$$

$$v_f' = \frac{1}{2} \cdot \frac{4}{3} \pi (a - \sigma)^3 + \frac{1}{2} \cdot \frac{4}{3} \pi (a - \sigma)^2 (a - \sigma(1 - \beta)) \qquad (6.13b)$$

By noting that in hexagonal packing a surface molecule is surrounded by 9 molecules instead of the 12 for an interior molecule, the relation

$$\psi'(o) = \frac{9}{12}\,\psi(o) \tag{6.14}$$

is obtained. Here $\psi(o)$ is represented as the Lennard-Jones potential for a lattice with hexagonal packing. Then,

$$Z\psi(o) = 12\varepsilon[1.0109(N\sigma^3/V_s)^4 - 2.4090(N\sigma^3/V_s)^2] \tag{6.15}$$

where σ and ε are the distance and the energy characteristic of the system. With the aid of Eqs. (6.13), (6.14), and (6.15), Eq. (6.12) is written as

$$\gamma = \frac{2}{\sqrt{3}}\,\sigma^{-2}\left(\frac{N\sigma^3}{\sqrt{2}\,V_s}\right)^{2/3}$$

$$\times \left\{ \varepsilon\left[3.613\left(\frac{N\sigma^3}{V_s}\right)^2 - 1.516\left(\frac{N\sigma^3}{V_s}\right)^4\right] - kT \ln \frac{1 - 0.875\left(\dfrac{N\sigma^3}{\sqrt{2}\,V_s}\right)^{1/3}}{1 - \left(\dfrac{N\sigma^3}{\sqrt{2}\,V_s}\right)^{1/3}} \right\}$$

$$\tag{6.16}$$

In this derivation, $\beta = 0.25$ is the value found by Prigogine and Saraga [20] and it has been introduced into Eq. (6.16) where it fits the experimental surface tension values of various simple compounds. Some of the calculated results are shown in Table 6-7. If we define the reduced quantities, $kT/\varepsilon = T^*$, $V/N\sigma^3 = V^*$, and $(N\sigma^3/V_s)^2 = \alpha$, the reduced surface tension γ^* is given by Eq. (6.17):

$$\gamma^* = \gamma\sigma^2/\varepsilon = [1/(V^*)^2](1.94 - 0.638\,T^*) \tag{6.17}$$

where the value $\alpha^{-1/2} = 1.01$, which is the average value of $V_s/N\sigma^3$ for Ne, A, CH_4, and N_2, has been introduced.

For the numerical calculation, however, we need the volume of a liquid as a function of temperature. For this function, the formula found empirically by Guggenheim [21] is used.

In Figure 6-3, the γ^* values from Eq. (6.17) are compared with those from other theories. It may be seen that the significant structure approach is most successful. The non-superposability of the values for He, H_2, and Ne is due to a quantum effect. Although the one-layer method is only an approximate one, it yields useful results and the calculations are simple. However, the Ree-Ree-Eyring equation includes the Lennard-Jones (6-12) potential, whose characteristic constants are only available for simple liquids. This fact prevents wide application of the theory. Recently, Lu, Jhon, Ree, and Eyring

Table 6-7 Comparison of Calculated Surface Tensions with Experimental Values

	ε/k (°K)	σ (Å)	V_s (cc/mole)	T (°K)	ρ (g/cc)	γ_{calc} (dyne/cm)	γ_{exp} (dyne/cm)
Argon	119.8	3.405	24.98	85.0	1.417	15.7	13.2
				90.0	1.385	14.7	11.9
Nitrogen	95.05	3.698	29.31	70.0	0.8446	11.0	10.5
				75.0	0.8217	10.2	9.39
				80.0	0.7988	9.37	8.27
				85.0	0.7759	8.60	7.20
				90.0	0.7530	7.87	6.16
Oxygen	118	3.46	23.77	70.0	1.239	18.2	18.3
				75.0	1.210	17.1	17.0
				80.0	1.191	16.2	15.7
				85.0	1.167	15.2	14.5
				90.0	1.143	14.3	13.2
Methane	148.2	3.817	30.94	93.2	0.449	16.5	18.0
				103.2	0.4362	15.0	15.8
				113.2	0.4218	13.5	13.7

[1c] derived another simple equation for surface tension using the Einstein oscillator model instead of the Lennard-Jones potential. The derivation is analogous to that of Ree, Ree, and Eyring, but this new equation is successfully applied not only to simple liquids but also to polar liquids and molten metals. Their estimation of the surface tension also yields reasonable agreement with experiment. These methods have been tested on the following systems.

Simple liquids:
    argon, neon, nitrogen, oxygen, chlorine, fluorine
polyatomic liquids:
    methane, carbon tetrachloride, phosphine, hydrogen bromide, ammonia, chloroform, chloromethane, benzene, p-xylene, m-xylene
liquid metals:
    Na, K, Ag, Au, Al, In, Pb, Mg, Zn, Hg, Tl.

6.6 SURFACE TENSION OF LIQUID MIXTURES

Several attempts were made to apply statistical mechanics to the calculation of surface tension of mixtures. Guggenheim [22] used a lattice model to describe the liquid state and restricted his considerations to mixtures of molecules of nearly the same size. Englert-Chwoles and Prigogine [23]

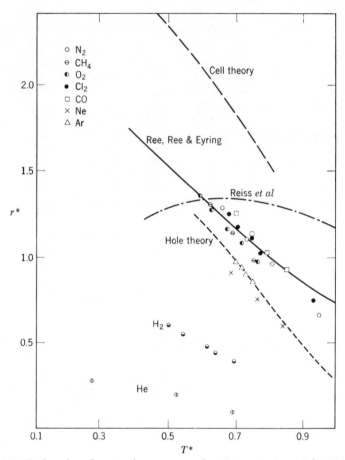

Figure 6-3 Reduced surface tension versus reduced temperature. (after Ree *et al.* [1b])

calculated the equilibrium surface tension of a mixture as a perturbation on the dynamic surface tension using a smoothed potential model. Unfortunately, they never calculated the surface mole fraction explicitly. Recently, Eckert and Prausnitz [24] derived an equation for the surface tension of a nonpolar liquid mixture from the application of the grand partition function to a cell model. The significant structure theory has been successfully applied to the calculation of the surface tensions of a number of pure liquids. Its extension to binary liquid mixtures appears feasible. At the present time, Kim, Jhon, Ree, and Eyring [25] have developed the monolayer approximation for the calculation of the surface tension of cryogenic liquid mixtures by using the liquid mixture partition function of significant structure theory.

REFERENCES

[1] (a) S. Chang, T. Ree, H. Eyring, and I. Matzner in *Prog. in Intern. Research on Thermodynamics and Transport Properties*, p. 88, Maşi and Tsai, eds., Academic Press, New York, 1962.
(b) T. S. Ree, T. Ree, and H. Eyring, *J. Chem. Phys.*, **41**, 524 (1964).
(c) W. C. Lu, M. S. Jhon, T. Ree, and H. Eyring, *J. Chem. Phys.*, **46**, 1075 (1967).

[2] S. Onno and S. Kondo, *Handbuch der Physik*, Vol. X, "Structur der Flussigkeiten," p. 134, Flugge, ed., Springer Verlag, Berlin, 1960.

[3] C. V. Raman and L. A. Ramdas, *Phil. Mag.*, **3**, 220 (1927).

[4] N. K. Adam, *The Physics and Chemistry of Surfaces*, 2nd Ed., Oxford University Press, 1938.

[5] E. C. C. Baly and F. G. Donnan, *J. Chem. Soc. (London)*, **81**, 907 (1902): G. Rudorf, *Ann. Physik*, **29**, 751 (1909).

[6] (a) S. Park, H. Pak, and S. Chang, *J. Korean Chem. Soc.*, **8**, 183 (1964).
(b) M. S. Jhon and S. Chang, *J. Korean Chem. Soc.*, **8**, 65 (1964).
(c) T. R. Thomson, H. Eyring, and T. Ree, *J. Phys. Chem.*, **67**, 2701 (1963).

[7] J. Grosh, M. S. Jhon, T. Ree, and H. Eyring, *Proc. Natl. Acad. Sci. (U.S.)*, **54**, 1419 (1965).

[8] K. C. Kim, W. C. Lu, T. Ree, and H. Eyring, *Proc. Natl. Acad. Sci. (U.S.)*, **57**, 861 (1967).

[9] D. Henderson, H. Eyring, and D. Felix, *J. Phys. Chem.*, **66**, 1128 (1962).

[10] (a) M. S. Jhon, J. Grosh, and H. Eyring, *J. Phys. Chem.*, **71**, 2253 (1967).
(b) R. Schmidt, M. S. Jhon, and H. Eyring, *Proc. Natl. Acad. Sci. (U.S.)*, **60**, 387 (1968).

[11] M. S. Jhon, G. Clemena, and E. R. Van Artsdalen, *J. Phys. Chem.*, **72**, 4155 (1968).

[12] M. S. Jhon, J. Grosh, and H. Eyring, unpublished work.

[13] W. Paik and S. Chang, *J. Korean Chem. Soc.*, **8**, 29 (1964).

[14] E. J. Fuller, "Significant Liquid Structure," Ph.D. Thesis, University of Utah, 1960.

[15] M. S. Jhon and H. Eyring, to be published.

[16] E. A. Moelwyn-Hughes, *Physical Chemistry*, 2nd. revised ed., Pergamon, New York, 1961.

[17] (a) H. Pak and S. Chang, *J. Korean Chem. Soc.*, **8**, 121 (1964).
(b) M. S. Jhon, E. R. Van Artsdalen, J. Grosh, and H. Eyring, *J. Chem. Phys.*, **47**, 2231 (1967).

[18] H. B. Lee, M. S. Jhon, and S. Chang, *J. Korean Chem. Soc.*, **8**, 179 (1964).

[19] C. V. Raman and L. A. Ramdas, *Phil. Mag.*, (7) 3, 220 (1920).

[20] I. Prigogine and L. Saraga, *J. Chim. Phys.*, **49**, 399 (1952).

[21] E. A. Guggenheim, *J. Chem. Phys.*, **13**, 253 (1945).

[22] Guggenheim, E. A., *Mixtures*, Oxford University Press, London, 1952, Chap. 9.

[23] A. Englert-Chwoles and I. Prigogine, *J. Chim. Phys.*, **55**, 16 (1958); A. Englert-Chwoles and I. Prigogine, *Nuovo Cimento*, **9**, Suppl. 1, p. 347 (1958).

[24] C. A. Eckert and J. M. Prausnitz, *A.I.Ch.E. Journal*, **10**, 677 (1964).

[25] S. W. Kim, M. S. Jhon, T. Ree, and H. Eyring, *Proc. Natl. Acad. Sci. (U.S.)*, **59**, 336 (1968).

chapter 7

THE DIELECTRIC
CONSTANT OF LIQUIDS

7.1 INTRODUCTION

The study of dielectric properties is a useful tool for understanding the structure of matter. The dielectric constant ε of a medium may be defined as the ratio of a field strength in a vacuum to that in the medium for the same distribution of charge. It may also be defined, and is measured, as the ratio of the capacitance C of a condenser filled with the material to the capacitance C_0 of the empty condenser, that is,

$$\varepsilon = \frac{C}{C_0} \tag{7.1}$$

The field intensity, D, perpendicular to the plate in a parallel empty condenser is given by

$$D = 4\pi\sigma \tag{7.2}$$

Here σ is the charge density of the surface at the plate and D is called the "electric displacement vector." If the homogeneous material with the dielectric constant ε is filled between the plates of the parallel condenser, the field strength E will decrease as follows:

$$E = \frac{4\pi\sigma}{\varepsilon} \tag{7.3}$$

Equation (7.3) is rewritten as

$$D = \varepsilon E \tag{7.4}$$

Then

$$D - E = 4\pi\sigma\left(1 - \frac{1}{\varepsilon}\right) = 4\pi\sigma\,\frac{\varepsilon - 1}{\varepsilon} \tag{7.5}$$

110

From the definition,

$$P = \sigma \frac{\varepsilon - 1}{\varepsilon} \qquad (7.6)$$

Eq. (7.5) can be rewritten as

$$D = E + 4\pi P \qquad (7.7)$$

Here P is called "the polarization." Substituting Eq. (7.4) into Eq. (7.7), we obtain

$$\frac{\varepsilon - 1}{4\pi} = \frac{P}{E} \qquad (7.8)$$

This is the fundamental electrostatic equation. If the polarization of the dielectric P and the field intensity E are evaluated, one can formulate the equation of the dielectric constant from Eq. (7.8). The polarization P has the following explicit form

$$P = \sum_i n_i \alpha_i F \qquad (7.9)$$

where n_i is the number of molecules of the ith species per cc, α_i is the polarizability of the ith species, and F is the local field strength. In early days (1879) Clausius and Mosotti assumed the Lorentz field which is given by

$$F = \frac{\varepsilon + 2}{3} E \qquad (7.10)$$

and derived the famous Clausius-Mosotti equation [1] using Eq. (7.8) and Eq. (7.9)

$$\frac{\varepsilon - 1}{\varepsilon + 2} \frac{M}{d} = \frac{4}{3} \pi N \alpha \qquad (7.11)$$

Here, M is the molecular weight, d is the density, N is Avogadro's number, and α is the polarizability.

Equation (7.11) is quite satisfactory for comparing the calculated dielectric constants of nonpolar liquids with the experimental values. However, the deviation of the calculated values from the experimental values for polar molecules is very large. In the polar liquid, besides the above mentioned static polarizability, there is an orientational polarizability which arises from orientation of the polar molecule in the field. The additional polarizability $\mu^2/3kT$ was introduced into Eq. (7.11) by Debye [2] giving,

$$\frac{\varepsilon - 1}{\varepsilon + 2} \frac{M}{d} = \frac{4}{3} \pi N \left(\alpha + \frac{\mu^2}{3kT} \right) \qquad (7.12)$$

In this derivation, the assumption was made that the interaction between two molecules with dipole moment μ is negligible. This equation leads to excellent predictions of the dielectric constants of polar molecules in the dilute gas phase or in dilute solution. It yields erroneous results, however, for concentrated polar liquids.

7.2 EARLY THEORIES OF POLAR LIQUIDS

Debye [3] modified his equation in an attempt to apply it to liquids in which dipole interactions are important. In polar liquids, each molecule is not allowed to rotate freely, but is restricted in its rotation by its neighbors. Debye assumed a potential energy of interaction, U, of the form,

$$U = -W \cos \theta$$

where θ is the angle between the axis of the permanent dipole moment and the instantaneous axis fixed by the surroundings, and W is the energy required to twist a molecule suddenly at right angles to its instantaneous axis when no external field is applied. This assumption leads to the following result.

$$\frac{\varepsilon - 1}{\varepsilon + 2} \frac{M}{d} = \frac{4}{3} \pi N \left(\alpha + \frac{\mu^2}{3kT} (1 - L^2(y)) \right) \tag{7.13}$$

where $L(y)$ is the Langevin function which is given by $L(y) = \coth y - 1/y$, and $y = W/kT$.

To account for the polarization in liquid water, y was taken as 11, and in the case of nitrobenzene, $y = 10$. Unfortunately, this model predicts much shorter relaxation times than are observed. On the other hand, Onsager [4] set out to explain the deviation in polar liquids from the original Debye equation by evaluating the local field precisely giving up the Lorentz field and returning to the fundamental electrostatic equations. The field F', surrounding a spherical vacuum in a homogeneous medium of dielectric constant ε_1 is

$$F' = \frac{3\varepsilon}{2\varepsilon_1 + 1} E \tag{7.14}$$

Onsager called this the cavity field. In addition to the cavity field, the dipole in a cavity is influenced by the electrostatic force of the surrounding molecules. This is called the reaction field. These fields are taken into account in evaluating the local field. In particular, on introduction of the internal dielectric constant, n^2, and assuming spherical, isotropically polarizable molecules, he obtained the equation:

$$\frac{(\varepsilon - n^2)(2\varepsilon + n^2)}{3\varepsilon} = 4\pi \frac{N}{V} \left(\frac{\mu^2}{3kT} \right) \left(\frac{n^2 + 2}{3} \right)^2 \tag{7.15}$$

where n is the refractive index appropriate to electronic and atomic polarizability, N is Avogadro's number, V is the molar volume of the liquid at temperature T, k is the Boltzmann constant, and μ the value of the permanent dipole moment of the isolated molecule. Equation (7.15) gives satisfactory results for many polar liquids, but gives poor agreement with experiment for an H-bonded liquid such as water.

In his derivation of Eq. (7.15), Onsager neglected short-range order. Kirkwood [5] attempted to account for the short-range directional interactions by introducing a correlation factor between a central molecule and its nearest neighbors. By using statistical mechanical methods with some assumptions, he derived the relation

$$\frac{(\varepsilon - n^2)(2\varepsilon + n^2)}{9\varepsilon} = \frac{4}{3}\pi\frac{N}{V}\left(\alpha + \frac{g\mu^2}{3kT}\right) \qquad (7.16)$$

Here g is called the Kirkwood correlation parameter.

7.3   THE DOMAIN THEORY OF THE DIELECTRIC CONSTANT OF H-BONDED LIQUIDS

Very recently, Hobbs, Jhon, and Eyring [6] proposed a domain theory of H-bonded liquids introducing significant structure concepts, and applied the theory successfully to water, and various forms of ice. The model proposed differs from that of Onsager and Kirkwood in an essential respect. This difference is in the method of taking account of short range order. In Onsager's model, this polarizability,

$$\frac{\mu^2}{3kT}\left(\frac{n^2+2}{3}\right)^2$$

is obtained by space-averaging $\mu\cos\theta$ over all possible values of the angle θ, which assumes that the relative population of dipole vectors is not constrained by short range directional interactions. Kirkwood considered the short range directional interaction in terms of the correlation factor, but again it was assumed that the central dipole was free to orient in a local field, F.

The above assumptions failed to explain [6] the residual entropy values which for ice at low temperatures has the value 0.870 e.u. as observed by Giaugue [7]. This entropy should persist only somewhat diminished in a hydrogen-bonded liquid like water. According to the proposed domain theory, the following assumptions are made.

(1) Liquid water, and ice, are made up of a mosaic of roughly brick-like domains with the dipoles in a particular domain having an average resultant moment, $\mu\cos\theta$, along the direction of maximum polarization for the domain,

while the direction of maximum polarization of neighboring domains tends to be rotated through 180° with respect to the first domain in the same way that magnets juxtapose south poles against north poles. For perfect tetrahedral bonding cos θ gives the value of unity, while for bent hydrogen bonds cos θ is correspondingly smaller. Figure 7-1 is helpful in understanding the geometry of this picture.

(2) In an electric field, the directions of maximum polarization of domains tend to orient in the field until all domains are either lined up with the field

Figure 7-1 This model of the structure of ice was kindly supplied to us by Dr. Melvin E. Zandler. The small balls represent hydrogen atoms, while the big balls represent oxygen atoms. We see that the direction of every dipole points in the same direction—outward and down. Thus, the projection of each dipole μ cos θ along the direction of maximum polarization for such a domain, is μ. (after Hobbs, Jhon, and Eyring. [6])

or against it. This orientation is presumably accomplished not so much by rotation of the domains as by the growth, one molecule at a time, of favorably oriented domains. Those favorably oriented domains continue to grow at the expense of the less favorably oriented domains until a steady state is reached. The relaxation process involves only rotation of single molecules at the interface between domains. The lack of sufficient degeneracy, as shown by the entropy [7], supports this local relaxation assumption. The resulting mean dipole moment for solid-like molecules is then

$$\mu = \frac{\mu \cos \theta e^{\mu \cos \theta F/kT} - \mu \cos \theta e^{-\mu \cos \theta F/kT}}{e^{\mu \cos \theta F/kT} + e^{-\mu \cos \theta F/kT}}$$

$$= \frac{\mu^2 \cos^2 \theta F}{kT} = \frac{\mu^2 GF}{kT} \quad \text{provided} \quad \frac{\mu \cos \theta F}{kT} \ll 1 \qquad (7.17)$$

Here F is the local field and $\cos^2 \theta = G$. Thus, for the solid-like part of water, the factor $\mu^2/3kT$ in Eq. (7.15) must be replaced by $\mu^2 G/kT$, while for the gas-like portion, the factor $\mu^2/3kT$ is retained since such molecules orient freely in the local field, F.

(3) The interface between domains in the mosaic is made up of Bjerrum faults, vacancies, and related mismatch structures. In the light of the foregoing discussion and from the significant structure theory of the liquid, the fraction V_s/V of the molecules of a liquid has solid-like properties and the remaining fraction $(V - V_s)/V$ has gas-like properties. The dielectric constant of water is then given by

$$\frac{(\varepsilon - n^2)(2\varepsilon + n^2)}{3\varepsilon} = 4\pi \frac{N}{V} \left(\frac{n^2 + 2}{3}\right)^2 \left(\frac{V_s}{V} \frac{\mu^2 G}{kT} + \frac{V - V_s}{V} \frac{\mu^2}{3kT}\right) \qquad (7.18)$$

For convenience in actual calculations, Eq. (18) may be rearranged to give

$$\varepsilon = \frac{n^2}{2} + \frac{n^4}{2\varepsilon} + 6\pi NG\left(\frac{\mu^2}{k}\right)\left(\frac{n^2 + 2}{3}\right)^2 \left(\frac{1}{VT}\right)\left[\frac{V_s}{V}\left(1 - \frac{1}{3G}\right) + \frac{1}{3G}\right] \qquad (7.19)$$

The only adjustable parameter, G, can be determined by evaluation at a single temperature using one experimental value of ε in Eq. (7.19). In the calculation, the value of G is taken as 0.964 for light and heavy water. The fact that the values found for G are nearly equal to unity is a very satisfying result. The small deviation of the value of G from unity reflects, at least in part, the deviation from tetrahedral structures due to bending of hydrogen bonds. For comparison, the data by Oster and Kirkwood are shown in Table 7-1 and the calculated values of the dielectric constant of light and heavy water are compared in Table 7-2 with experiment. Judging from the data in Table 7-1 for the small temperature interval involved, the Kirkwood correlation parameter gives a poorer agreement with the temperature dependence

Table 7-1 Dielectric Constant
of Water as Calculated by
Oster and Kirkwood [5]

$T (°K)$	ε_{calc}	ε_{obs}
273	84.2	88.0
298	78.2	78.5
335	72.5	66.1
356	67.5	59.5

Table 7-2 Dielectric Constants of Light and Heavy Water
According to Significant Structure Theory

$T (°K)$	ε_{obs}	ε_{calc}	V (cc)	V_s (cc)	n^2
			H_2O		
273	88	88	18.0	17.85	1.78
373	56	60	18.8	17.7	1.74
473	35	38	20.8	17.7	1.64
573	20	22	25.3	17.7	1.51
			D_2O		
278	85.8	85.3	18.11	17.88	1.78
293	80.1	80.9	18.13	17.85	1.78
313	73.1	75.4	18.20	17.85	1.77
333	66.7	69.8	18.36	17.85	1.76

than that given by Eq. (7.19). Also, the Oster-Kirkwood model, which assumes free rotation about the hydrogen bond of the nearest neighbors of a central molecule, seems less appropriate for water than the model of water as a mosaic of domains. Clathrate structures that have been proposed for pure water such as the dodecahedra suggested by Pauling would yield dielectric constants that are much too low.

7.4 RELAXATION TIMES OF WATER AND ICE

According to Debye [2], the dielectric constant ε_2 is a function of frequency and satisfies the equation

$$\varepsilon_2 = (\varepsilon_0 - \varepsilon_\infty) \frac{\omega\tau}{1 + \omega^2\tau^2} \tag{7.20}$$

Here from the absolute rate theory [8], τ is given as

$$\frac{1}{2\tau} = k' = \frac{kT}{h} \, e^{-\Delta G^{\ddagger}/RT} \qquad (7.21)$$

and ε_0 and ε_∞ are the dielectric constants at low and high frequencies, and ω is the circular frequency of the applied potential. Substituting the values [9] of $\tau_{H_2O} = 5.3 \times 10^{-6} \exp(13{,}250/RT)$ for ordinary ice and $\tau_{D_2O} = 7.7 \times 10^{-6} \exp(13{,}400/RT)$ for solid deuterium oxide into Eq. (7.21), we find $k'_{H_2O} = 9.4 \times 10^4 \exp(-13{,}250/RT)$ for ordinary ice and $k'_{D_2O} = 6.5 \times 10^4 \exp(-13{,}400/RT)$ for solid deuterium oxide, respectively. The estimated value for the frequency factor $(kT/h) \exp \Delta S^{\ddagger}/R$ for a simple unimolecular rate constant where $\Delta S^{\ddagger}$ is small is 10^{13}. If we divide 9.4×10^4 by 10^{13}, we obtain 9.4×10^{-9} for the fraction, f, of molecules which rotate from one domain into a neighboring domain better-oriented with respect to the field. The linear dimension of such a domain would be about a tenth of a micron if we assumed that only one molecule at a time in the entire interface between domains can migrate to the other domain. If the entropy of activation $\Delta S^{\ddagger}$ is negative, the domain is correspondingly smaller.

If the reaction coordinate is rotation about an axis normal to the axis of the dipole, the frequency factor would be larger for H_2O than D_2O by a factor of $\sqrt{2}$. The observed frequency factors are consistent with this requirement. Manson, Cagle, and Eyring [10] found 13,800 calories for the activation energy for the rate of growth of microcrystals of ice from supersaturated vapor to visible size. This correspondence in activation energies for crystal growth and passage between domains in ice is an interesting result. Manson, Cagle, and Eyring [10] concluded that liquid domains of about 40 water molecules must nucleate as a unit to form the bridge required in crystal growth. Jhon et al. [11] successfully explained the thermodynamic transport and surface properties of water on the assumption that it contained domains of about 46 molecules which changed their structure as a unit from an "ice I-like" to an "ice III-like" state. This indicates, as one would expect, that domains in water are much smaller than those in ice. Because of the abundance of fluidized vacancies in water, liquid domains lack long-range order, and we expect that the relaxation time for water will have an apparent entropy of relaxation much nearer unity and an activation energy less by the amount of energy to form a vacancy, i.e., about 10,000 calories. This is what is found experimentally. Thus, $\tau = 9 \times 10^{-11}$ is reported by Collie, Hasted, and Ritson [12] for the relaxation time of water. The domain theory thus seems in reasonable accord with the observed dielectric constants of light and heavy ice and water and also in accord with the observed relaxation times.

7.5 THE DIELECTRIC CONSTANT OF ALCOHOLS AND OF THE SUPERCRITICAL REGION OF WATER

Here, we examine the applicability [13] of domain theory to the supercritical regions of steam for which no satisfactory theoretical studies appear to have been made. And also, as a further application of Eq. (7.19), the calculated values of the dielectric constants of various lower aliphatic alcohols are compared with experiment. For the calculation of the dielectric constant of steam at supercritical temperatures under high pressure, we have to consider the pressure effect on Eq. (7.19).

In a compressed region of dense gas or liquid, the pressure effect [14] on V_s is not negligible, since some solids are fairly compressible. Thus, the pressure dependence of the solid volume is

$$V_{sp} = V_s(1 - \beta \Delta P) \tag{4.27}$$

Here V_{sp} indicates the solid volume under pressure, β is the solid compressibility, ΔP is the pressure minus the vapor pressure at the melting temperature. If ΔP is not large, V_{sp} is equal to V_s, since β for ice is 12×10^{-6} atm^{-1}. But at pressures in excess of a few hundred atmospheres V_{sp} is noticeably changed. The values [6] of $G = 0.964$ and $\mu = 1.84D$ and the experimental values of approximately 250 atm was used for the pressure. The index of refraction was estimated in a similar way [6] assuming the validity of the Clausius-Mosotti theorem. The results are summarized in Table 7-3, and show

Table 7-3   The Dielectric Constant of Steam at Various Densities

d (g/cm³)	V (cc)	n^2	$T = 661°K$		$T = 651°K$	
			ε_{calc}	ε_{obs}	ε_{calc}	ε_{obs}
0.1	180.16	1.064	1.81	1.83	1.83	1.85
0.2	90.08	1.129	3.09	2.91	3.12	2.97
0.3	60.05	1.198	4.89	—	4.95	—
0.4	45.04	1.270	7.22	7.12	7.31	7.27
0.5	36.03	1.344	10.07	7.72	10.16	9.94

good agreement with the observed values [15]. As a further application of Eq. (7.19), the dielectric constants of various lower aliphatic alcohols were calculated. The single parametric values of G used [11b] are 1.202 for methanol; 1.237 for ethanol; 1.312 for n-propanol; 1.238 for isopropanol. Values of G, slightly larger than unity, can be analyzed as follows. In liquid alcohol systems, we may assume that there is some degree of polymerization of the alcohol molecules. If polymerization occurs, and the p molecules orient as a unit, the mean average dipole moment of molecules will be $(p\mu)^2/p = p\mu^2$. Presumably it becomes easier to form the polymer as the alcohols increase in length, and it is increasingly difficult to build the polymer as the side groups become larger and more bulky. The values found are consistent with this point of view. Exemplifying the model the dielectric constants of liquid methanol are calculated over a wide temperature range, and are listed in Table 7-4. The results are quite satisfactory.

Table 7-4 Dielectric Constant of Methanol

$T(°K)$	ε_{obs}	ε_{calc}	V (cc)	n^2
298	32.70	32.71	40.48	1.769
333	26.53	26.85	42.41	1.725
373	20.66	21.41	44.87	1.677
473	9.35	10.51	57.94	1.499
503	6.31	6.81	72.65	1.385

7.6 DIELECTRIC CONSTANT OF LIQUID MIXTURES

Here, we describe the extension [13] of the domain theory to mixtures. According to Hobbs, Jhon, and Eyring, the dielectric constant of a hydrogen-bonded liquid is given by

$$\frac{(\varepsilon_A - n_A^2)(2\varepsilon_A + n_A^2)}{3\varepsilon_A} = 4\pi \frac{N}{V_A} \left(\frac{n_A^2 + 2}{3}\right)^2 \left(\frac{V_{sA}}{V_A} P_{sA} + \frac{V_A - V_{sA}}{V_A} P_{gA}\right) \quad (7.22)$$

where

$$P_{sA} = \frac{\mu_A^2 G_A}{kT} \quad \text{and} \quad P_{gA} = \frac{\mu_A^2}{3kT}$$

The expression for a second hydrogen-bonded liquid B is obtained by replacing the subscript A by B in Eq. (7.22). Analogously, the dielectric constants ε_M of mixtures of liquid A and B should be given by

$$\frac{(\varepsilon_M - n_M^2)(2\varepsilon_M + n_M^2)}{3\varepsilon_M} = 4\pi \frac{N}{V_M} \left(\frac{n_M^2 + 2}{3}\right)^2 \left(\frac{V_{sM}}{V_M} P_{sM} + \frac{V_M - V_{sM}}{V_M} P_{gM}\right) \quad (7.23)$$

Here, the letters with the subscript M represent the quantities for the mixture. In our calculations, n_M^2 and V_M are taken from the references indicated. For the evaluation of the value of P_{gM}, we neglected the interaction between A and B molecules in the gas-like degrees of freedom. Then,

$$P_{gM} = \frac{\mu_A^2}{3kT} X_A + \frac{\mu_B^2}{3kT} X_B \tag{7.24}$$

However, we have to consider the A-B type interaction for the calculation of P_{sM} for the solid-like degree of freedom. If we represent the mole fractions of each component by X_A and X_B, the probabilities of the occurrence of A-A, B-B, and A-B contacts of molecules are given by X_A^2, X_B^2, and $2X_A X_B$ respectively under the condition of random mixing.

$$P_{sM} = X_A^2 P_{sA} + X_B^2 P_{sB} + 2X_A X_B P_{sAB} \tag{7.25}$$

Here P_{sAB} can be defined as

$$\sqrt{P_{sA} P_{sB}} \tag{7.26a}$$

or

$$\frac{P_{sA} + P_{sB}}{2} \tag{7.26b}$$

We used the following relation for V_{sM}

$$V_{sM} = V_{sA} X_A + V_{sB} X_B \tag{7.27}$$

Substituting Eqs. (7.24), (7.25), (7.26a), and (7.27) into Eq. (7.23), and rearranging for convenience, we obtain the following:

$$\varepsilon_M = \frac{n_M^2}{2} + \frac{n_M^4}{2\varepsilon_M} + \frac{3}{2} 4\pi \frac{N}{V_M} \left(\frac{n_M^2 + 2}{3} \right)^2$$

$$\times \left[\frac{V_{sA} X_A + V_{sB} X_B}{V_M} \left(X_A^2 \frac{\mu_A^2 G_A}{kT} + X_B^2 \frac{\mu_B^2 G_B}{kT} + 2X_A X_B \frac{\sqrt{G_A G_B} \mu_A \mu_B}{kT} \right) \right.$$

$$\left. + \frac{V_M - V_{sA} X_A - V_{sB} X_B}{V_M} \left(\frac{\mu_A^2}{3kT} X_A + \frac{\mu_B^2}{3kT} X_B \right) \right] \tag{7.28}$$

Equation (7.28) is tested by applying it to the typical polar mixture of water with alcohol.

    Using the values appearing in References 6, 11b, 16, and 17 the dielectric constants of the water-methanol system at several temperatures and compositions varying from 0 % to 100 % are compared with experiment in Table 7-5. The calculated results are in excellent agreement with experiment [16–18] over the entire composition range. The applicability of Eq. (7.28) is further tested for the water-dioxane systems. The study of the dielectric constant of this system is important in distinguishing between various theories and in interpreting experiments reported in recent years [11a].

Table 7-5 Dielectric Constant of the Water-Methanol System in Various Proportions

Mole % (MeOH)	283 (°K)		298 (°K)		313 (°K)	
	ε_{calc}	ε_{obs}	ε_{calc}	ε_{obs}	ε_{calc}	ε_{obs}
0	83.02	83.83	78.52	78.48	74.18	73.15
0.2432	—	—	62.72	65.55	—	—
0.2725	63.79	66.05	—	—	55.84	56.20
0.4575	54.09	56.20	50.40	51.67	46.96	47.40
0.8350	39.86	41.55	35.89	37.91	34.12	34.60
1.0000	34.75	35.75	32.68	32.61	30.37	29.80

The addition of dioxane to water should change the liquid structure and, consequently, the dielectric behavior. Recently, Smyth *et al.* [19] found that this system has two relaxation times, in contrast to the single relaxation time found for water. Two relaxations can be understood in terms of the domain model: One relaxation is the rapid rotation of individual molecules at the interface between pure water domains; the second is the rotation of islands of p water molecules giving a second relaxation time. The addition of dioxane to water tends to increase the number of small islands of water.

For the estimation of P_{sAB}, we use Eq. (7.26b) since the dipole moment of the dioxane in liquid is negligibly small. It should be noted that P_{sA}^* in an A-B contact is different from P_{sA} in an A-A contact, since the former involves the simultaneous rotation of the p molecules. Substituting Eqs. (7.24), (7.25), (7.26b), and (7.27) into Eq. (7.23), we obtain:

$$\varepsilon_M = \frac{n_M^2}{2} + \frac{n_M^4}{2\varepsilon_M} + \frac{3}{2} 4\pi \frac{N}{V_M}$$

$$\times \left[\frac{V_{sA} X_A + V_{sB} X_B}{V_M} \left(X_A^2 \frac{\mu_A^2 G_A}{kT} + X_A X_B \frac{\mu_A^2 G_A^*}{kT} \right) \right.$$

$$\left. + \frac{V_M - V_{sA} X_A - V_{sB} X_B}{V_M} \left(\frac{\mu_A^2}{3kT} X_A \right) \right] \left(\frac{n_M^2 + 2}{3} \right)^2 \quad (7.29)$$

In the calculation, the values [13] of $G_A = 0.964$, $G_A^* = 3.47$, $\mu_A = 1.84$, $V_{sA} = 17.65$ cc and $V_{sB} = 76.84$ cc were used.

The calculated results of the dielectric constants of the water-dioxane mixtures are compared with experiment [20] in Table 7-6. Calculation and experiment are in good agreement. The only significant deviation which occurs in the range of high concentration of the dioxane can be explained by considering that G_A^* may decrease due to depolymerization of the water in high concentrations of dioxane.

Table 7-6 The Dielectric Constants of Water-Dioxane Systems of Various Compositions

Mole % $(C_4H_8O_2)$	V (cc)		n		ε			
					298 (°K)		313 (°K)	
	298 (°K)	313 (°K)	298 (°K)	313 (°K)	Calc.	Obs.	Calc.	Obs.
0	18.08	18.17	1.3325	1.3309	78.61	78.5	74.03	73.12
20	30.97	31.31	1.3869	1.3820	59.72	60.4	55.01	56.26
40	44.47	45.00	1.4035	1.3978	41.92	42.9	38.80	39.54
60	58.24	59.03	1.4127	1.4050	27.04	25.9	24.91	23.72
80	72.03	73.13	1.4173	1.4095	13.76	10.7	12.68	9.91
100	85.83	87.23	1.4204	1.4126	2.02	2.101	1.995	2.098

REFERENCES

[1] O. F. Mosotti, *Mem. di mathem efiscadi Modena*, **24**, 2.49 (1850).
     R. Clausius, *Die mechanische Wärmelehre*, **2**, 62–97 (Vieweg, 1879).
[2] P. Debye, *Polar Molecules*, Chemical Catalog Co., New York, 1929.
[3] P. Debye and W. Ramm, *Ann. der Physik*, **28**, 28 (1937).
[4] L. Onsager, *J. Am. Chem. Soc.*, **58**, 1486 (1936).
     See also *Statistical Mechanics and Dynamics* by Eyring, Henderson, Stover, and Eyring (John Wiley and Sons, 1964, page 213) for a simple discussion.
[5] (a) J. G. Kirkwood, *J. Chem. Phys.*, **7**, 911 (1939).
     (b) G. Oster and J. G. Kirkwood, *J. Chem. Phys.*, **11**, 175 (1943).
[6] M. E. Hobbs, M. S. Jhon, and H. Eyring, *Proc. Natl. Acad. Sci. (U.S.)*, **56**, 31 (1966).
[7] W. F. Giauque and M. F. Ashley, *Phys., Rev.*, **43**, 81 (1933).
[8] H. Eyring, D. Henderson, B. J. Stover, and E. M. Eyring, *Statistical Mechanics and Dynamics*, Wiley, New York, 1964, p. 227.
[9] R. P. Auty and R. H. Cole, *J. Chem. Phys.*, **20**, 1309 (1952).
[10] J. E. Manson, F. W. Cagle, Jr., and H. Eyring, *Proc. Natl. Acad. Sci. (U.S.)*, **44**, 156 (1958).
[11] (a) M. S. Jhon, J. Grosh, T. Ree, and H. Eyring, *J. Chem. Phys.*, **44**, 1465 (1966).
     (b) M. S. Jhon, E. R. Van Artsdalen, J. Grosh, and H. Eyring, *J. Chem. Phys.*, **47**, 2231 (1967).

[12] C. H. Collie, J. B. Hasted, and D. M. Ritson, *Proc. Phys. Soc.* (*London*), **60**, 145 (1948).

[13] M. S. Jhon and H. Eyring, Debye Memorial Issue, *J. Am. Chem. Soc.*, **90**, 3071 (1968).

[14] T. S. Ree, T. Ree, and H. Eyring, *Proc. Natl. Acad. Sci.* (*U.S.*), **48**, 501 (1962).

[15] J. K. Fogo, S. W. Benson, and C. S. Copeland, *J. Chem. Phys.*, **27**, 212 (1954).

[16] *Chemical Physics of Ionic Solutions*, ed. B. E. Conway and R. G. Barradas, Wiley, New York, 1966, p. 211.

[17] C. Carr and J. A. Riddick, *Ind. Eng. Chem.*, **43**, 692 (1951).

[18] B. E. Conway, *Electrochemical Data*, Elsevier, New York, 1952.

[19] (a) C. J. Clemett, E. Forest, and C. P. Smyth, *J. Chem. Phys.*, **40**, 2123 (1964).
    (b) S. K. Garg, and C. P. Smyth, *J. Chem. Phys.*, **43**, 2959 (1965).

[20] J. Timmermans, *Physicochemical Constants of Binary System in Concentrated Solutions*, Vol. IV, Interscience, New York, 1959.

chapter 8

TWO-DIMENSIONAL LIQUIDS AND SOLID ADSORBANTS

8.1 TWO-DIMENSIONAL LIQUIDS

Very few theories of two-dimensional liquids have been developed. Devonshire [1] extended the Lennard-Jones and Devonshire cell theory of liquids to the two-dimensional case and concluded that the critical temperature of the two-dimensional fluid was about half of that of the bulk liquid. Recently, Wang, Ree, Ree, and Eyring [2] applied the significant structure theory of liquids to describe a two-dimensional liquid of hard disks. The system of hard circular disks is a convenient and meaningful test of the two-dimensional liquid theory, since Alder and Wainwright [3] published exact data for this system obtained from molecular dynamics calculations. The scaled particles theory [4] also predicts quite accurate results for circular disks. According to the significant structure theory, the partition function for a liquid is as follows:

$$f = [f_s]^{V_s N/V} [f_g]^{\frac{V-V_s}{V}\,N} \qquad (3.2)$$

here

$$f_s = f_{\text{solid}}(\text{Einstein})[1 + n_h \exp\left(-\varepsilon_h/RT\right)] \qquad (3.7)'$$

For a two-dimensional liquid, we write the partition function as:

$$f' = \{f'_{\text{solid}}(\text{Einstein})[1 + n'_h \exp\left(-\varepsilon'_h/RT\right)]\}^{A_s N/A} \{f'_g\}^{\frac{(A-A_s)N}{A}} \qquad (8.1)$$

where every quantity is defined analogously to the quantities in Eq. (3.2) and Eq. (3.7)'; the prime is added to differentiate the two-dimensional quantities from their three-dimensional counterparts; A_s is the molar area of the two-dimensional solid at its melting point, and A is the molar area of the two-dimensional liquid. Henderson [5] extended the earlier calculations of significant structure theory by using the Lennard-Jones and Devonshire cell model instead of the Einstein oscillator model for the solid-like degrees of

124

freedom in the formulation of the partition function, and applied this modified partition function to a system of hard spheres.

If a perfect-gas partition function for f_g' and a Lennard-Jones and Devonshire partition function for f_{solid}' are used, then Eq. (8.1) becomes

$$f' = \left\{ \frac{2\pi mkT}{h^2} a_f \exp\left(-\frac{Z\phi(o)}{2kT}\right)[1 + n_h' \exp\left(-\varepsilon_h'/RT\right)] \right\}^{A_sN/A}$$

$$\times \left\{ \frac{2\pi mkT}{h^2} \frac{eA}{N} \right\}^{\frac{(A-A_s)N}{A}} \quad (8.2)$$

Here, Z is the number of nearest neighbors and is equal to 6 for close packing: a_f is the "free area" of a molecule and is given by

$$a_f = \int_{\text{cell}} \exp\left\{ -\frac{Z[\phi(r) - \phi(o)]}{2kT} \right\} dr, \quad (8.3a)$$

where $\phi(o)$ is the potential energy when a molecule of the system is at its two-dimensional cell center.

For a system of hard circular disks, the following relations are true:

$$\phi(r) = 0 \qquad r > \sigma$$
$$\qquad \infty \qquad r < \sigma \qquad (8.3b)$$

where σ is the diameter of hard disks, and the $\varepsilon_h' = 0$ relation comes from the equation $\varepsilon_h' = a'\phi(o)A_s/(A - A_s)$ according to the significant structure theory.

Using the above-mentioned relation Eq. (8.2) is simplified to

$$f' = \left[\frac{2\pi mkT}{h^2} \frac{eA}{N}\right]^N \left[a_f(1 + n_h')\frac{N}{eA}\right]^{N\left(\frac{A_s}{A}\right)} \quad (8.4)$$

Alder et al. [3b] obtained the relation $A_s = (\frac{4}{3})A_0$ (A_0 being the close packed area) from the study of the computer-generated oscilloscope traces of disk trajectories in the two-phase region. Accordingly, we use $A_s = 1.312A_0$ and the free area $a_f = (a - \sigma)^2 = 0.05552a^2$ [2] in the following calculation. According to the significant structure theory, n_h' is given as $n(A - A_s)/A_s$ where n is nearly equal to the coordination number 6 for close packing. We can now rewrite Eq. (8.4):

$$f' = \left[\frac{2\pi mkT}{h^2} \frac{eA}{N}\right]^N \left[0.02358\left(6 - 5\frac{A_s}{A}\right)\right]^{A_sN/A} \quad (8.5)$$

The pressure of the two-dimensional system, π, is given by

$$\pi = kT\left(\frac{\partial \ln f'}{\partial A}\right)_T \quad (8.6)$$

or

$$\frac{\pi A}{RT} = 1 - 1.312 \frac{A_0}{A} \ln \left[0.02358 \left(6 - 6.560 \frac{A_0}{A} \right) \right]$$

$$+ \left(1.312 \frac{A_0}{A} \right)^2 \frac{5}{6 - 6.560(A_0/A)} \quad (8.7)$$

In deriving Eq. (8.7) the relation $A_s = 1.312 \, A_0$ has been introduced. The calculated results are compared with those of machine calculations by Alder and Wainwright in Figure 8-1. The virial expansion of Eq. (8.7) is

$$\frac{\pi A}{RT} = 1 + 1.433 \frac{b}{A} + 0.872 \left(\frac{b}{A} \right)^2 + 0.394 \left(\frac{b}{A} \right)^3 + 0.211 \left(\frac{b}{A} \right)^4 + 0.119 \left(\frac{b}{A} \right)^5 + \cdots$$

$$(8.8)$$

compared with the exact virial expansion [6]

$$\frac{\pi A}{RT} = 1 + \frac{b}{A} + 0.7820 \left(\frac{b}{A} \right)^2 + 0.5324 \left(\frac{b}{A} \right)^3 + 0.3338 \left(\frac{b}{A} \right)^4 + 0.1992 \left(\frac{b}{A} \right)^5 + \cdots$$

$$(8.9)$$

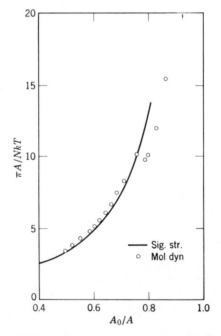

Figure 8-1 Compressibility factor for a system of rigid discs (two-dimensional liquid). (after Wang *et al.* [2])

where $b = (\pi/2)\sigma^2 N$. The agreement is only fair. From the relation $S = [\partial(kT \ln f)/\partial T]_V$, the excess entropy S^E, over that of a perfect gas at the same temperature and pressure is readily shown to be

$$\frac{S^E}{R} = \ln \frac{\pi A}{RT} + 1.312 \frac{A_0}{A} \ln \left[0.02358 \left(6 - 6.56 \frac{A_0}{A} \right) \right] \tag{8.10}$$

In deriving Eq. (8.10), the partition function for an ideal two-dimensional gas has been taken as

$$f_g'(\text{ideal}) = \left[(2\pi m kT/h^2) \left(\frac{eA_{id}}{N} \right) \right]^N \tag{8.11}$$

and the relation $\pi A_{id}/RT = 1$ has been substituted. On the other hand, the excess entropy is calculated by using the following equation [3]:

$$\frac{S^E}{R} = \ln \frac{\pi A}{RT} + \int_\infty^{A/A_0} \left[\frac{\pi A}{RT} - 1 \right] \frac{A_0}{A} d\left(\frac{A}{A_0} \right) \tag{8.12}$$

Figure 8-2 shows the excess entropy found by the significant structure approach, and by molecular dynamics calculations.

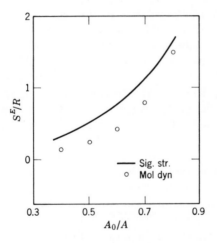

Figure 8-2 Excess entropy for a system of rigid discs. (after Wang *et al.* [2])

In Table 8-1, the excess entropies and the compressibility factors obtained by the molecular dynamics method (MD), the significant structure theory (SST), the scaled-particle theory [4] (SPT), and the cell theory (CT) are compared.

Table 8-1 Comparison of the Thermodynamic Data for a Hard Disk·Liquid
Calculated from Various Theories with the Machine Calculations

A_0/A	S^E/R				$\pi A/RT$			
	MD	SST	CT	SPT	MD	SST	CT	SPT
0.4	0.13	0.33	0.59	0.12	2.49	2.75	2.85	2.46
0.5	0.24	0.53	0.86	0.23	3.41	3.61	3.51	3.35
0.6	0.42	0.81	1.15	0.41	5.03	4.90	4.53	4.81
0.7	0.80	1.18	1.47	0.73	7.73	7.15	6.57	7.50
0.8	1.51	1.73	1.95	1.35	11.99	12.58	9.53	13.28

8.2 THE THEORY OF SUBMONOLAYER PHYSICAL ADSORPTION

In the vicinity of the boundary between solid and gas phases, it is usually
true that the surface density is several orders of magnitude greater than the
gas phase value. Thus, it is convenient to think of the region in the immediate
vicinity of the surface as a separate adsorbed phase. If the forces holding
molecules near the surface are of a physical nature (e.g., Van der Waals), it is
called "physical adsorption." There are several approaches to a study of the
problem of adsorption of gases on solids. One can start with partition functions describing the adsorbed and bulk phases and equate the chemical
potential of the two phases or one can apply the Gibb's adsorption isotherm
to a model equation of state. The significant structure approach belongs to
the former category. Recently, McAlpin and Pierotti [7] applied the significant
structure theory of liquids to a simplified system whose surface is a structure-
less uniform plane and with physical adsorption occurring in the submono-
layer region.

According to their model, the adsorbed phase is a two-dimensional fluid
adsorbed on a solid energetically homogeneous plane surface. The molecules
of the fluid interact with the solid and vibrate normal to the surface of the
solid. The interaction is assumed to fall off rapidly enough with distance [8]
to ensure that, at low temperatures, only the first layer nearest the surface
will be populated. Preserving the form of Eqs. (3.7) and (3.2), we can write
the partition function for a two-dimensional adsorbed fluid as

$$f_{\text{ads}} = \{f_{2s}(1 + n_h \exp(-\varepsilon_a/RT))\}^{NA_s/A} \cdot \{f_{2g}\}^{N \frac{(A - A_s)}{A}} \tag{8.13}$$

where f_{ads}, f_{2s}, $(1 + n_h \exp(-\varepsilon_a/RT))$, and f_{2g} are respectively the partition
function for the adsorbed phase, for the two-dimensional solid-like structure
(including the adsorbate-adsorbent interaction terms), the positional de-

generacy, and the partition function for the two-dimensional gas-like structure (including the adsorbate-adsorbent interaction terms); A_s is the molar area of the solid adsorbate; A is the molar area of the adsorbed phase at a given surface coverage θ; and N is the total number of molecules in the adsorbed phase.

Now we consider the relation,

$$A = \frac{A_m^0}{\theta}$$

where A_m^0 is equal to LA_m/N_m, A_m is the area of the surface, and N_m is the number of molecules on the surface at $\theta = 1$, respectively, and L is Avogadro's number. Substituting this relation into Eq. (8.13) and recalling that $N = \theta N_m$, we obtain the result:

$$f_{ads} = \{f_{2s}(1 + n_h \exp(-\varepsilon_a/RT))\}^{\theta^2 N_m A_s/A_m^0}\{f_{2g}\}^{\theta N_m(1-\theta A_s/A_m^0)} \quad (8.14)$$

In actual calculations of the isotherm, the isosteric heat of adsorption, the critical properties, and the virial coefficient, the following relation will be used. The Helmholtz free energy for the adsorbed phase is given by

$$F_{ads} = -kT \ln f_{ads} \quad (8.15)$$

and the chemical potential for the adsorbed phase and the gas phase are

$$\mu_{ads} = \left(\frac{\partial F_{ads}}{\partial N}\right)_{A_m, T} \quad (8.16a)$$

and

$$\mu_{gas} = \mu_{gas}^0 + kT \ln P \quad (8.16b)$$

Equating (8.16a) and (8.16b), an isotherm is obtained. The isosteric heat of adsorption q_{st} is given by

$$q_{st} = RT^2\left(\frac{\partial \ln P}{\partial T}\right)_\theta \quad (8.17)$$

The critical coverage, θ_c, and the critical temperature T_c can be determined using the following conditions:

$$\left(\frac{\partial P}{\partial \theta}\right)_{T, N_m} = 0, \quad \left(\frac{\partial^2 P}{\partial \theta^2}\right)_{T, N_m} = 0 \quad (8.18)$$

The critical pressure can be obtained from θ_c, T_c, and the isotherm. Finally, the second virial coefficient B_2 has the value:

$$B_2 = \left[\frac{\partial[\pi A/RT]}{\partial(1/A)}\right]_{1/A \to 0} \quad \text{here} \quad \pi = -\left(\frac{\partial F_{ads}}{\partial A}\right)_T \quad (8.19)$$

8.3   APPLICATION OF THE THEORY TO THE INERT GASES

The theory has been tested for argon and krypton adsorbed on graphite surfaces. For liquid argon or krypton, f_{2s} in Eq. (8.14) can be expressed as the product of the partition function of a molecule in a two-dimensional Einstein crystal (with θ_E as the Einstein temperature), and a partition function for the vibration normal to the surface. Then,

$$f_{2s} = \left(\frac{\exp(-\theta_E/2T)}{1 - \exp(-\theta_E/T)} \right)^2 \cdot \exp \frac{U_0 + W}{RT} \cdot \left(\frac{\exp(-h\nu_0/2kT)}{1 - \exp(-h\nu_0/kT)} \right) \quad (8.20)$$

Here U_0, W, and ν_0 are the molar adsorbate-adsorbent interaction energy, the molar two-dimensional lattice energy of the adsorbate, and the frequency of the vibration.

The two-dimensional gas-like partition function, f_{2g}, is given by

$$f_{2g} = \left(\frac{2\pi mkT}{h^2} \right) \frac{eA}{L} \cdot \exp \frac{U_0}{RT} \cdot \left(\frac{\exp(-h\nu_0/2kT)}{1 - \exp(-h\nu_0/kT)} \right) \quad (8.21)$$

As pointed out in Chapter 3, ε_a is an inverse function of the fraction of nearest-neighbor vacancies. In normal liquids, this fraction is approximately 0.1, whereas in the adsorbed phase (even at θ as high as 0.7), the fraction of nearest-neighbor vacancies is three times as high as in a normal liquid. Therefore, it is assumed that ε_a is negligible in the calculation. If $\theta = 1$ the state is taken to be a two-dimensional ideal solid, and A_s is equal to A_m^0 from the relation $A = A_m^0/\theta$. Using this result in Eq. (8.14) and combining Eqs. (8.20) and (8.21), we find the partition function for the adsorbed phase becomes

$$f_{\text{ads}} = \{B[1 + Z(1 - \theta)]\}^{\theta^2 N_m} \left(\frac{C}{\theta} \right)^{\theta N_m (1 - \theta)} \quad (8.22)$$

where B and C are defined as follows:

$$B = D \left(\frac{\exp(-\theta_E/2T)}{1 - \exp(-\theta_E/T)} \right)^2 \exp \left(\frac{W}{RT} \right) \quad (8.23)$$

$$C = D \frac{2\pi mkT}{h^2} \frac{eA_m^0}{L} \quad (8.24)$$

and

$$D = \left(\frac{\exp(-h\nu_0/2kT)}{1 - \exp(-h\nu_0/kT)} \right) \exp \left(\frac{U_0}{RT} \right) \quad (8.25)$$

Using Eqs. (8.15), (8.16a), (8.16b), and (8.17), the adsorption isotherm and the isosteric heat of adsorption q_{st} are given by

$$\ln P = -\frac{\mu_{gas}^0}{kT} - 2\theta \ln \left\{ \frac{B\theta}{C} [1 + Z(1 - \theta)] \right\}$$

$$- \theta^2 \left(\frac{1 + Z - 2Z\theta}{[1 + Z(1 - \theta)]\theta} \right) - \ln \left(\frac{C}{\theta} \right) + 1 \quad (8.26)$$

and

$$q_{st}(\theta \cdot T) = \left[2W - 2R\theta_E + 2RT - 4R\theta_E \left(\frac{\exp(-\theta_E/T)}{1 - \exp(-\theta_E/T)} \right) \right] \theta$$

$$+ \left\{ U_0 - RT \left[\left(\frac{h\nu_0}{2kT} \right) \left(\frac{1 + \exp(-h\nu_0/kT)}{1 - \exp(-h\nu_0/kT)} \right) \right] + \frac{3}{2} RT \right\} \quad (8.27)$$

The conditions of Eq. (8.18) give the following critical values:

$$\theta_c = 0.3834$$

$$T_c = \left(\frac{4.93 \times 10^9}{MA_m^0} \right) \cdot \left(\frac{\exp[(W - R\theta_E)/RT_c]}{[1 - \exp(-\theta_E/T_c)]^2} \right) \quad (8.28)$$

where θ_c is the critical coverage, T_c the critical temperature, and M the molecular weight of the adsorbate.

Before comparing the theoretical and experimental isotherms, a few words should be said concerning the evaluation of W, the two-dimensional lattice energy, and about θ_E, the Einstein characteristic temperature, as well as certain other quantities. The two-dimensional lattice energy W was obtained by summing the Sinanoglu and Pitzer modification [9] of the Lennard-Jones (6-12) potential over the nearest 18 neighbors in the hexagonal lattice and integrating over the rest of the lattice. The Einstein characteristic temperature θ_E was obtained by multiplying the ratio of the perturbed and unperturbed frequencies (determined from the curvature of the potential energy curves) by the Einstein temperature of the bulk adsorbate. Quantities such as U_0, W, ν_0, and A_m^0 are obtainable from the references given. Table 8-2 contains the values of the molecular properties used. Theoretical adsorption isotherms for argon and krypton adsorbed on graphite were calculated using Eq. (8.26) and compared with experiment [10] in Figures 8-3 and 8-4. The agreement is excellent at low and intermediate coverage. Theoretical and experimental [11] isotherms for Kr on graphite at 77.8°K are compared in Figure 8-5. The agreement is again very good, but not as good as with argon. This may be due to the few percent error in W, which was calculated as described above. For krypton there was no experimental verification of the third order dispersion effect as there was in the case of argon. The theoretical isosteric

Table 8-2   Molecular Parameters and Constants for Ar and Kr
on Graphite

	Ar	Kr
U_0 (cal/mole)	2220	2860
v_0 (sec^{-1})	1.20×10^{12}	1.00×10^{12}
A_m^0/L (Å^2/molecule)	14.8 (77.6°K)	19.5 (78°K)
	15.4 (90.1°K)	
ε/k (°K)	119.8	171
σ (cm)	3.405	3.60
W (cal/mole)	710	890
θ_E (°K)	45.0	33.8
V_m (cc (S.T.P)/g)	3.66	2.86

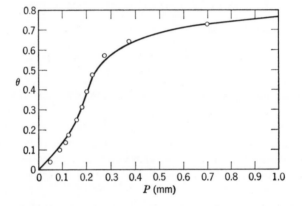

Figure 8-3   Argon-graphite isotherm at 77.6°K for Ar on P-33. Solid line is theoretical. Points are data from Ref. 10. (after McAlpin and Pierotti [7])

heats of adsorption of argon on graphite at 84°K and 150°K were calculated
using Eq. (8.27) and in Figure 8-6 are compared with experimental heats
determined by Ross and Oliver [10] for Ar on P-33 at 84°K. Agreement is
quite good. The slope of the experimental curve at 84°K is 1000 cal/mole while
the predicted value is 1050 cal/mole. Sams, Constabaris, and Halsey [12]
reported a slope for this system of 810 cal/mole, while the slope of the
theoretical curve at 150°K is 833 cal/mole. According to the present model,
θ_c is found to be 0.3824, whereas that of a two-dimensional van der Waals
fluid is 0.333 and that of a localized Bragg-Williams monolayer is 0.5. The
critical temperature for argon on graphite is predicted to be 65°K whereas
the experimental value is around 68°K [13]. The predicted T_c for Kr on

graphite is 79°K. There is no reliable experimental T_c to compare with, but it appears to be around 82°K [14].

Thus, the significant structure model for physical adsorption gives good agreement between theory and experiment for many properties. It is not possible to obtain two-dimensional virial coefficients from the present model. In order to obtain second virial coefficients, the appropriate modification of Eyring's expression, $1 + n((V - V_s)/V_s)e^{-\varepsilon_h/RT}$, for the three-dimensional liquid is required. If ε is taken equal to zero, this expression becomes $1 + n(1 - \theta)/\theta$ for a two-dimensional liquid. But the results with $\varepsilon_h = 0$ are not satisfactory.

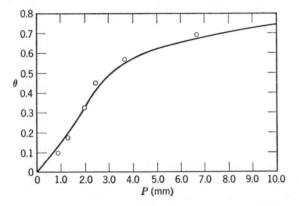

Figure 8-4 Argon-graphite isotherm at 90.1°K for Ar on P-33. Solid line is theoretical. Points are data from Ref. 10. (after McAlpin and Pierotti [7])

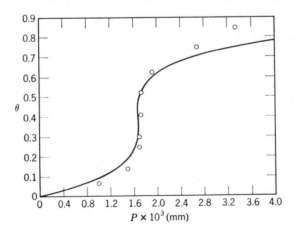

Figure 8-5 Krypton-graphite isotherm at 77.8°K for Kr on P-33. Solid line is theoretical. Points are data from Ref. 11 (after McAlpin and Pierotti [7])

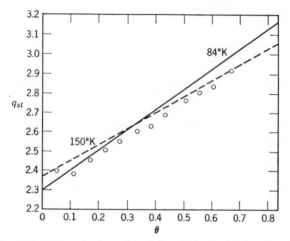

Figure 8-6 Isosteric heats of adsorption (kilocalories/mole) for Ar on P-33 versus coverage for argon on graphite. The lines are theoretical. The data points are from Ref. 10.

Extension of Ree, Ree, and Eyring's [15] dense gas theory should be useful, since it gives better predictions of the second virial coefficient than other available theories do. Very recently, Schmidt and Jhon [16] have assumed that the adsorbed phase occurring in physical adsorption is actually a dense gas. The preliminary results obtained are encouraging.

8.4 APPLICATION OF THE THEORY TO POLYATOMIC GASES AND GAS MIXTURES

If the model discussed here is to be of practical use, it should be applicable to more complex molecules than monatomic gases and gas mixtures. The results so far obtained are encouraging.

Experimental studies of adsorption at low coverage which show mono-layer adsorption on homogeneous surface have been reported for H_2, O_2, CO_2, and hydrocarbons. These molecules involve rotational and vibrational partition functions, in addition to the simple partition function for adsorbed atoms. The data [17] indicate that for the larger hydrocarbons, there is a decrease in rotational entropy of 2–4 cal/mole deg because of hindered rotation. Hindered rotation is an important aspect of the theory of physical adsorption of the above systems and is especially significant in chromatographic separations at low temperature [18]. The model for adsorption of single gases can be extended to the calculation of the physical adsorption of mixed gases [19] using analogous forms of the mixture partition function discussed in Chapter 4.

REFERENCES

[1] A. F. Devonshire, *Proc. Roy. Soc. (London)*, **A163**, 132 (1937).
[2] Y. L. Wang, T. Ree, T. S. Ree, and H. Eyring, *J. Chem. Phys.*, **42**, 1926 (1965).
[3] (a) B. J. Alder and T. E. Wainwright, *J. Chem. Phys.*, **33**, 1439 (1960).
(b) B. J. Alder, W. G. Hoover, and T. E. Wainwright, *Phys. Rev., Letters*, **11**, 241 (1963).
[4] H. L. Frisch, *Advan. Chem. Phys.*, **6**, 229–289 (1964).
[5] D. Henderson, *J. Chem. Phys.*, **39**, 1857 (1963)
[6] F. H, Ree and W. G. Hoover, *J. Chem. Phys.*, **40**, 939 (1964).
[7] (a) J. J. McAlpin and R. A. Pierotti, *J. Chem. Phys.*, **41**, 68 (1964).
(b) J. J. McAlpin, Ph.D. Thesis, Georgia Institute of Technology, 1963.
[8] R. A. Pierotti, *J. Chem. Phys.*, **36**, 2515 (1962
[9] O. Sinanoglu and K. S. Pitzer, *J. Chem. Phys.*, **32**, 1279 (1960).
[10] S. Ross and J. P. Oliver, *J. Phys. Chem.*, **65**, 608 (1961).
[11] S. Ross and W. Winkler, *J. Colloid Sci.*, **10**, 330 (1955).
[12] J. R. Sams, G. Constabaris, and G. D. Halsey, *J. Phys. Chem.*, **66**, 2154 (1962).
[13] C. F. Prenzlow and G. D. Halsey, *J. Phys. Chem.*, **61**, 1158 (1957).
[14] R. A. Pierotti, *J. Phys. Chem.*, **66**. 1810 (1962).
[15] T. S. Ree, T. Ree, and H. Eyring, *Proc. Natl. Acad. Sci. (U.S.)*, **48**, 501 (1962).
[16] R. Schmidt and M. S. Jhon unpublished work.
[17] S. E. Hoory and J. M. Prausnitz, *Trans. Faraday Soc.*, **63**, Part 2, 455 (1967).
[18] E. M. Mortensen and H. Eyring, *J. Phy. Chem.*, **64**, 433 (1960).
[19] D. M. Young and A. D. Crowell, *Physical Adsorption of Gases*, Butterworths, Washington, 1962, Chap. 11.

chapter 9

LIQUID THEORY AND REACTION KINETICS

9.1 INTRODUCTION

We next consider the general formulation of the theory of absolute reaction rates in the liquid state. This will be followed by a discussion of its applicability. Using the theory of liquid mixtures, the chemical potentials of the various components can be calculated. The chemical potentials of the liquid reactants are, of course, required to calculate rates of reaction.

9.2 FORMULATION OF ABSOLUTE REACTION RATE THEORY [1]

Reaction rates can in general be written in terms of elementary processes whose rates are characterized by a single potential barrier. For the net rate of the forward crossing of the potential barrier in or out of the liquid state, we have

$$\text{rate} = \kappa \, \frac{kT}{h} \, (C_f^\ddagger - C_b^\ddagger) \tag{9.1}$$

where $C_f^\ddagger$ is the concentration of activated complexes per length, $\delta = h(2\pi m^\ddagger kT)^{-1/2}$, which would be in equilibrium with reactants, while $C_b^\ddagger$ represents the analogous quantity for activated complexes in equilibrium with products. Here, κ, k, T, and h are, as usual, the transmission coefficient, the Boltzmann constant, the absolute temperature, and the Planck constant, respectively. We represent by $m^\ddagger$ the mass of the activated complex for motion across the potential energy barrier.

The potential energy change along the reaction coordinate for an elementary reaction is shown schematically in Figure 9-1. Formation of the activated complex may be indicated by the following equation:

$$aA + bB + \cdots \rightarrow C^\ddagger + dD \rightarrow \cdots \leftarrow lL + mM + \cdots \tag{9.2}$$

where $C^\ddagger$ is the concentration of activated complexes and D is a possible intermediate.

136

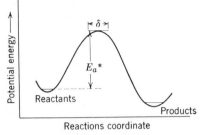

Figure 9-1 Schematic of the potential energy changes along the reaction coordinates. $E_a^{\ddagger}$ is the energy of activation at the absolute zero of temperature with the zero of energy taken as the ground state of the reactants.

At equilibrium, $C_f^{\ddagger}$ is equal to $C_b^{\ddagger}$, since reactants are colliding to form activated complexes with the same frequency that products are colliding to form the activated complexes moving back across the barrier. However, there is no interaction between the various activated complexes since their concentrations are extremely dilute and they are separated spatially. Consequently, if the products were suddenly removed, the equilibrium number of activated complexes moving in the forward direction would continue to be formed at substantially the same rate by reactants. On this basis, we can still use equilibrium statistics to relate reactants to their activated complexes, as well as classical statistics to describe the activated state in nonequilibrium systems.

The Boltzmann relation gives the ratio of the number of particles in their respective energy states at equilibrium:

$$\frac{n_i}{n_j} = \frac{\omega_i e^{-\varepsilon_i/kT}}{\omega_j e^{-\varepsilon_j/kT}} \tag{9.3}$$

where ε_i and ε_j are the energies of states i and j relative to a common reference level, and ω_i and ω_j are the degeneracies of the i and j states, respectively. Rewriting Eq. (9.3), and defining the absolute activity λ, we obtain

$$\frac{n_i}{\omega_i e^{-\varepsilon_i/kT}} = \frac{n_j}{\omega_j e^{-\varepsilon_j/kT}} \equiv \lambda$$

$$n_i = \lambda \omega_i e^{-\varepsilon_i/kT}$$

$$n_j = \lambda \omega_j e^{-\varepsilon_j/kT}$$

$$\vdots \qquad \vdots$$

$$\sum_i n_i = \lambda \sum_i \omega_i e^{-\varepsilon_i/kT} = N \tag{9.4}$$

We see that the quantity λ is the number of molecules N in the system divided by the partition function, f, for the system. From Eq. (9.4), we have

$$\lambda = \frac{N}{\sum_i \omega_i e^{-\varepsilon_i/kT}} = \frac{N}{f} \tag{9.5}$$

Equation (5) can be rewritten as follows:

$$\lambda = \frac{N}{f} = \frac{N/V}{f/V} = \frac{C}{F} \tag{9.6}$$

where C is the concentration and F is the partition function per unit volume. Substituting Eq. (9.6) into Eq. (9.1), we obtain

$$\text{rate} = \kappa\, \frac{kT}{h} (\lambda_{C_f}^{\ddagger} F_0^{\ddagger} - \lambda_{C_b}^{\ddagger} F_0^{\ddagger}) \tag{9.7}$$

where

$$F_0^{\ddagger} = e^{-E_a\ddagger/kT} F^{\ddagger} \tag{9.8}$$

Here $E_a^{\ddagger}$ is defined in Figure 9-1 and $F^{\ddagger}$ is the partition function per unit volume for the activated complex, $F^{\ddagger}$ being written with the ground state of the complex as the zero of energy. The absolute activity can be written in general [2] as

$$\lambda_i = e^{\mu_i/kT} \tag{9.9}$$

where

$$\mu_i = \left(\frac{\partial A}{\partial n_i}\right)_{T,V,n_j} = \left(\frac{\partial G}{\partial n_i}\right)_{T,P,n_j} \tag{9.10}$$

Here μ_i is called the chemical potential and A and G are the Helmholtz free energy and the Gibbs free energy, respectively. Using equilibrium theory to relate reactants to the activated complexes crossing the barrier in the forward direction gives

$$a\mu_A + b\mu_B + \cdots = \mu_{C_f}^{\ddagger} + d\mu_D + \cdots \tag{9.11}$$

which with Eq. (9.9) gives

$$\lambda_A^a \lambda_B^b \cdots = \lambda_{C_f}^{\ddagger} \lambda_D^d \cdots \tag{9.12}$$

In a similar manner, for the backward direction

$$\lambda_L^l \lambda_M^m \cdots = \lambda_{C_b}^{\ddagger} \lambda_D^d \cdots \tag{9.13}$$

Using Eqs. (9.12) and (9.13), we can rewrite Eq. (9.7) as follows:

$$\text{rate} = \kappa\, \frac{kT}{h} \frac{F_0^{\ddagger}}{\lambda_D^d} [(\lambda_A^a \lambda_B^b \cdots) - (\lambda_L^l \lambda_M^m \cdots)] \tag{9.14}$$

With the use of Eqs. (9.9) and (9.10), the rate of a reaction may take on the following useful forms:

$$\text{rate} = \kappa \frac{kT}{h} \frac{F_0^{\ddagger}}{\exp\left(\dfrac{d\mu_D + \cdots}{kT}\right)}$$

$$\times \left[\exp\left(\frac{a\mu_A + b\mu_B + \cdots}{kT}\right) - \exp\left(\frac{l\mu_L + m\mu_M + \cdots}{kT}\right)\right] \tag{9.15}$$

or

$$\text{rate} = \kappa \frac{kT}{h} \frac{F_0^{\ddagger}}{\exp\left[\dfrac{1}{kT}\left(d\dfrac{\partial A}{\partial n_D} + \cdots\right)\right]} \left\{\exp\left[\frac{1}{kT}\left(a\frac{\partial A}{\partial n_A} + b\frac{\partial A}{\partial n_B} + \cdots\right)\right]\right.$$

$$\left. - \exp\left[\frac{1}{kT}\left(l\frac{\partial A}{\partial n_L} + m\frac{\partial A}{\partial n_M} + \cdots\right)\right]\right\} \tag{9.16}$$

The rate expressed in terms of chemical potentials, Eq. (9.13), leads to Onsager's reciprocal relations [3][4] only in the limit of small displacements from equilibrium.

Equation (9.16), the rate equation expressed in terms of the partial of the Helmholtz free energy with respect to particle number, can be the complete expression for reactions in solution, if the appropriate partition functions for liquid mixtures are available. This construction of the partition function for the mixture from the significant structure theory of liquid will be described next.

9.3 EVALUATION OF THE ABSOLUTE ACTIVITY FROM THE MIXTURE PARTITION FUNCTION

For a system of N particles, the Helmholtz free energy A is given in terms of classical statistical mechanics as

$$A = -kT \ln\left[\frac{1}{N! \, h^{3N}} \int e^{-H/kT} \, d\omega\right] \tag{9.17}$$

H, the Hamiltonian, is the sum of the kinetic and potential energies, $d\omega$, that is, $3N$ coordinates, q_i, and $3N$ moments, p_i, are required to characterize the system. However, for liquid systems, gross simplifications are necessary to carry out the integration and this approach is given up in favor of a model approach. The most successful model so far is the significant liquid structure theory [5].

According to this theory, the following analogous mixture partition function [6][7][8] for the liquid has been developed (see Chapter 4).

$$f_{\text{mix}} = \frac{(N_1 + N_2)!}{N_1! N_2!} f_1 f_2 (1 + n_h' e^{-\varepsilon'/RT})^{N(V_s'/V)} \exp\left(N \frac{V_s'}{V} \frac{E_s'}{RT}\right) \quad (9.18)$$

where

$$f_i = f_{s_i}^{N_i(V_s'/V)} f_{g_i}^{N_i(V - V_s')/V}$$

$$V_s' = X_1 V_{s_1} + X_2 V_{s_2} + X_1 X_2 \delta \quad (9.19)$$

$$E_s' = X_1^2 E_{s_1} + X_2^2 E_{s_2} + 2 X_1 X_2 (1 + \xi)(E_{s_1} E_{s_2})^{1/2}$$

Here X_i are the mole fractions; V_{si} and E_{si} the respective molar volume of the solid and energy of sublimation of the ith component; and δ and ξ the factors characteristic of the particular binary mixture. The number of holes, n_h', for the mixture is written as

$$n_h' = n \frac{V - V_s'}{V_s'} = (X_1 n_1 + X_2 n_2) \frac{V - V_s'}{V_s'} \quad (9.20)$$

and

$$\varepsilon' = \frac{a' V_s' V_s'}{V - V_s'}$$

where

$$a' = X_1 a_1 + X_2 a_2$$

Then, the absolute activity of component 1 has the form

$$\lambda_1 = X_1 [f_1' f_2' (1 + n_h' e^{-\varepsilon'/RT})^{(V_{s1} + X_2^2 \delta)/V} e^{V_s' E_s'/VRT}]^{-1} e^{(V - V_s')/V} e^{X_2 Q} \quad (9.21)$$

where

$$f_1' = f_{s_1}^{\phi_s/V} f_{g_1}^{(V - \phi_s)/V} \quad (9.22)$$

$$f_2' = f_{s_2}^{\phi_s'/V} f_{g_2}^{-\phi_s/V} \quad (9.23)$$

and

$$\phi_s = [(X_1^2 + 2 X_1 X_2) V_{s_1} + X_2^2 V_{s_2} + 2 X_1 X_2^2 \delta]$$

$$\phi_s' = X_2^2 [(V_{s_1} + V_{s_2}) + (X_2 - X_1)\delta] \quad (9.24)$$

The Q in the exponent is a large sum of relatively simple terms and is a function of the terms in E_s', V_s', ε', and n_h'. As $X_1 \to 1$ and $X_2 \to 0$, the absolute activity for a component in a binary mixture, Eq. (9.21) approaches the absolute activity of the pure liquid.

Thus, with the use of the general formulation of reaction rate, Eq. (9.14), and of the method for obtaining the absolute activities for the liquid mixture, we can calculate the absolute reaction rates in solution, provided, of course, we also know the partition function for the activated complex.

REFERENCES

[1] H. Eyring and D. W. Urry, Berichte der Bunsengessellschaft für, *Phys. Chem. Bd.*, **67**, Nr. 8, 731 (1963).
[2] R. H. Fowler and E. A. Guggenheim, *Statistical Thermodynamics*, Cambridge University Press, 1939, p. 66.
[3] L. Onsager, *Phys. Rev.*, **37**, 405 (1931).
[4] L. Onsager, *Phys. Rev.*, **38**, 2265 (1931).
[5] H. Eyring and T. Ree, *Proc. Natl. Acad. Sci. (U.S.)*, **47**, 526 (1961).
[6] K. Liang, H. Eyring, and R. P. Marchi, *Proc. Natl. Acad. Sci. (U.S.)*, **52**, 1107 (1964).
[7] S. Ma, and H. Eyring, *J. Chem. Phys.*, **42**, 1920 (1965).
[8] B. A. Miner and H. Eyring, *Proc. Natl. Acad. Sci. (U.S.)*, **53**, 1227 (1965).

SUBJECT INDEX

AUTHOR INDEX